THE LITTLE BOOK OF EXOPLANETS

the

LITTLE BOOK

of

EXOPLANETS

JOSHUA WINN

PRINCETON UNIVERSITY PRESS *PRINCETON AND OXFORD*

Published by Princeton University Press
41 William Street, Princeton, New Jersey 08540
99 Banbury Road, Oxford OX2 6JX

press.princeton.edu

All Rights Reserved

ISBN 978-0-691-21547-1
ISBN (e-book) 978-0-691-22117-5

British Library Cataloging-in-Publication Data is available

Editorial: Abigail Johnson
Production Editorial: Karen Carter
Text and Jacket/Cover Design: Jessica Massabrook
Production: Jacqueline Poirier
Publicity: Kate Farquhar-Thomson and Matthew Taylor
Copyeditor: Karen Verde

This book has been composed in Bembo Std

Printed on acid-free paper. ∞

Printed in the United States of America

1 3 5 7 9 10 8 6 4 2

CONTENTS

In memory of my father, Martin Winn
(1940–2022)

THE LITTLE BOOK OF EXOPLANETS

CHAPTER ONE

INTRODUCTION

Based only on the fact that you have opened this book and are reading this page, I'll venture to guess we have something in common. At some point in your life, you were outside on a clear night looking at the stars, and you started wondering. Do any of those stars have planets? Are there other worlds like the Earth?

My earliest memory of this state of mind was during a childhood family vacation to the Grand Canyon. We arrived at our campsite in evening twilight. My father lifted me on top of our Chevrolet Impala where I could lie down and look up. The air was cold, but the hood was still warm from the long drive to the campground. It was the first time I'd ever seen a truly dark night sky. Bright stars began appearing in their familiar patterns—but here, far from city lights, they gleamed like diamonds on black velvet. As twilight deepened and my eyes adjusted to the darkness, it seemed like a fog was slowly dissipating, revealing multitudes of fainter stars. Unfamiliar and numerous, the faint stars were disorienting. I could barely make out the constellations anymore.

I knew from my astronomy books and magazines that each of those points of light was an enormous nuclear furnace with the same power and grandeur as the Sun, which had been baking us all day before it set below the horizon. With a seemingly infinite number of stars on display, it felt impossible that the Sun and the Earth were unique or special in any way. Was each one of those points of light somebody else's Sun? Which one was home to a young alien, lying atop its father's warm spaceship, looking toward me?

At that time, decades ago, there was not a single star in the sky that was known with certainty to have planets of its own. Now, thanks to advances in astronomical technology, thousands of planets have been detected around other stars—some of them around those very stars in the sky that you and I have gazed upon and wondered about. Aldebaran, an orange-red star in the constellation Taurus, has a planet at least six times more massive than Jupiter. Kochab, the brightest star in the bowl of the Little Dipper, has a giant planet, as does Pollux, the star at the head of one of the mythological twins in the constellation Gemini. Edasich, a star within the sinuous body of Draco the Dragon, has two planets so massive that we're not sure whether to call them planets. One of them travels so slowly around the star that it probably hasn't made it all the way around even once since my childhood trip to the Grand Canyon, four decades ago.

About two decades ago, I became an astronomer. Thanks to lucky timing, I was able to join the scientific journey from complete ignorance about planets outside the Solar System to

CHAPTER ONE

definitive evidence that they exist around at least 30% of Sun-like stars and strong evidence for an even higher abundance of planets. The newly discovered planets include potentially Earth-like worlds along with many exotic planets that bear little resemblance to any of the members of the Solar System. We've found planets on highly elongated orbits, planets on the brink of destruction by the gravitational force of a nearby star, planets as light and puffy as cotton candy, planets orbiting two stars at the same time, and planets that probably have oceans of lava. Some of the new planets were anticipated by authors of science fiction. Others have inspired new stories.

Planets that belong to stars other than the Sun are called *extrasolar planets*. Usually, the name is shortened to *exoplanets*—although increasingly, I just call them *planets*, and I think this will soon become common practice. After all, based on the statistics of our surveys of nearby stars, we can be sure that at least 99.99999999% of all the planets in the galaxy are orbiting stars other than the Sun. Doesn't this overwhelming majority deserve the general name *planet*, without any prefix? The infinitesimal minority of planets that happen to share our star should be called *solar planets* or (less seriously) *endoplanets*.

Questions about names also arise when we try to decide what types of astronomical bodies deserve the name "planet." Recent discoveries have led to disagreements. Should we impose a minimum or maximum mass, or a restricted range of orbital distances, for an object to qualify as a planet? What should we call a Jupiter-mass object that exists alone in the emptiness of space, far from any star? This book will expose and explain the discoveries that have led to these controversies, but without dwelling on debates over terminology. Arguments about the correct names for things are often needlessly controversial (is Pluto a planet?) and rarely as interesting as the things themselves. Sometimes, we get so caught up in debating, we forget how lucky we are to be alive in an age when the frontiers of knowledge expand so rapidly that our nomenclature needs time to catch up.

The discovery of exoplanets was one of humanity's longest-awaited scientific achievements. Twenty-five centuries ago, Greek philosophers speculated about the possibility of other worlds and whether the Earth is unique. Yet, it was only about 25 years ago that exoplanetary science began in earnest. The reason for the long delay was that exoplanets are (to put it mildly) difficult to detect. Planets are puny and faint by astronomical standards. Even with one of the world's best telescopes, a planet is easily lost in the glare of the star that it orbits. Trying to see an Earth-like exoplanet around a Sun-like star is like trying to spot a firefly while someone is pointing a powerful searchlight directly in your eyes.

This book is about how the obstacles to detecting exoplanets were overcome, and why the study of exoplanets has become one of the most exciting and rapidly advancing areas of science. Since the mid-1990s, the number of known exoplanets has grown nearly exponentially with time. When I was a graduate student, I could count the number of known exoplanets on my fingers. I knew their names and characteristics as if they were members of my household. These days, though, counting the known exoplanets requires more fingers (and toes) than are possessed by my family and friends. While it is no longer possible to be on a first-name basis with every exoplanet, it is now possible to conduct census-style statistical studies of exoplanets to obtain clues about their origin, composition, and fate.

One of my main goals in writing this book was to convey not only what we have learned, but also how we have

CHAPTER ONE

obtained this knowledge. For example, think again about the view of all those stars in the night sky. How do we *know* they are other Suns? Those pinpricks of light would seem to have little in common with the majestic orb that heats and illuminates our world. Even to consider the possibility that the stars are Suns required bold imagination. The ancient Greek philosophers Anaxagoras and Aristarchus were among those who suspected that stars are Sun-like bodies at remote distances, but the definitive evidence did not arrive until the nineteenth century, when astronomers managed to measure the distances to a few of the nearest stars. Before we knew the distances to any of the stars, it was impossible to say whether the stars are relatively feeble light sources located nearby, or if they emit just as much light as the Sun and appear faint only because they are much farther away. But how can the distance to a star be measured?

The answer is a geometrical technique that surveyors call *triangulation* and astronomers call *parallax*. To see parallax for yourself, extend one of your arms in front of your face and raise a finger. Close your left eye while you look toward your finger and the scene in the background. Now switch eyes: open your left eye and close your right eye. It looks like your finger suddenly shifted to the right. Switch from one eye to the other, and your finger shifts back and forth. This happens because your right eye views your finger from a different angle than your left eye, and sees your finger projected in front of a different part of the background scene. If you know the distance between your eyes, and you measure the shift in the apparent position of your finger—that's the *parallax*—then you can use trigonometry to calculate the distance to your finger.

When surveyors use this technique, they don't blink their eyes. They use telescopes to make two sightings of a distant

6

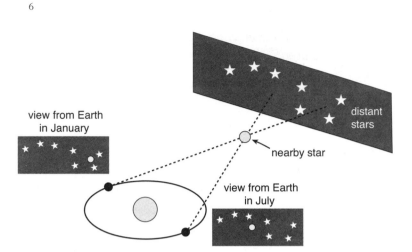

view from Earth
in January

distant
stars

nearby star

view from Earth
in July

FIGURE 1.1. As the Earth orbits the Sun, our changing perspective makes a nearby star seem to move relative to more distant stars, an effect called parallax. In this illustration, the effect is exaggerated—the angular shift is smaller than 0.001° even for the nearest stars.

object from two different locations. When astronomers use this technique, we take advantage of the Earth's motion around the Sun. We take a picture of a nearby star, making sure the image also includes distant background stars in the same part of the sky. Then we wait six months for the Earth to go halfway around its orbit, and take another picture (figure 1.1). The nearby star plays the same role as your finger: it appears to shift in position relative to the background stars.

In practice, we're never satisfied with a single pair of images—we obtain images as continuously as possible over the course of several years. In the resulting collection of images, a nearby star drifts slowly relative to the background stars, because of the star's actual motion through the galaxy, while also appearing to wiggle back and forth once per year, because of the parallax effect. By combining measurements of the parallax-induced motion and our knowledge of the diameter of the Earth's orbit, we can calculate the distances to the stars.

Parallax was one of the keys that unlocked the vault of secrets about the true nature of stars. Aristarchus was well versed in geometry and knew the importance of measuring stellar parallaxes, but he and his contemporaries had no way to measure the positions of stars with the necessary precision. To achieve that goal would require two millennia of technological progress, culminating in 1838, when Friedrich Bessel announced he had measured the parallax of a double star called 61 Cygni.[1] His results, and those of subsequent astronomers, confirmed that even the nearest stars are hundreds of thousands of times farther from the Earth than the Sun. If the Sun were placed at the same distance as 61 Cygni, it would appear to be a typical star, just barely qualifying as one of the hundred brightest stars in the sky. The similarity between the brightness of the Sun and the stars, after correcting for distance, was recognized as evidence that they are similar physical entities.

The distances to the stars are staggering. Even Alpha Centauri, the nearest star system visible to the naked eye, is approximately 40,000,000,000,000 kilometers away. To cope with such a mind-boggling number, it helps to use scientific notation and write it as 4×10^{13} kilometers. (The exponent on the 10 tells us how many zeros to write after the 4.) It also helps to express the distance as 4.4 light-years, implying that the starlight's journey from Alpha Centauri to the Solar System takes 4.4 years. Pondering these vast distances makes the

---

1. Out of necessity, I'm moving swiftly through history. Suggestions for further reading on this topic and others can be found at the end of this book.

Solar System seem like a one-horse town. The distance between Earth and Mars at their closest approach is about 55,000,000 kilometers, which sounds like a lot, but is only 3 light-minutes. That's close enough to be within reach of our rockets, and indeed, we have sent spacecraft to explore all the planets in the Solar System. Our probes have penetrated the clouds of Venus, taken road trips around Mars, skimmed across Saturn's rings, and snapped pictures of Pluto.

To explore exoplanets in the same fashion might require another few millennia of technological advancement. Our fastest current rockets would need tens of thousands of years to reach Alpha Centauri. To shorten the trip duration to fit within a human lifetime will require a gigantic leap in our capabilities. The members of an initiative called Breakthrough Starshot, funded by venture capitalist (and ex-physicist) Yuri Milner, are developing a concept for a probe that could reach Alpha Centauri only 30 years after launch. The probe would not achieve high speed from the thrust of a rocket. Instead, it would be pushed from behind by high-power laser beams fired from Earth's surface. Even in this ingenious and audacious scheme, to attain a high enough acceleration, the probe could be no more massive than *one gram*. That would not allow for much luggage, and there would be no overhead compartments.

So, for now, it is a sobering fact that up-close inspection of exoplanets is out of the question. We cannot collect rocks from the Kochab system, breathe the air of Arcturan worlds, or plunder the planet of Pollux. Exoplanets are detached from human activity, with the sole exception of what we can learn via telescopes, peering from afar.

Fortunately, telescope technology keeps advancing with no sign of letting up. Although we're stuck in the Solar System, we've learned to extract gigabytes of data from the trickles of light that reach us from distant stars and planets. We can focus starlight into the sharpest images allowed by the laws of optics; we can measure changes in brightness with a precision of five decimal places; we can spread starlight into a rainbow and measure the intensity of each of a hundred thousand colors. Whenever I think about how much we have learned about the universe despite being trapped on a tiny rocky outpost in an arbitrary location, I am filled with admiration for the collective power of our civilization and with gratitude that many aspects of the universe have proven to be comprehensible by our primate brains.

A major theme of this book is that we can learn a lot about an exoplanet even when we cannot see it in an image. Instead of relying on images, we can use indirect techniques based on our ability to perform precise measurements of starlight and our knowledge of the laws of physics. For example, in the constellation of Aquarius, at the top of the water-bearer's pail, is a dusky red star—much too faint to see by eye—which until recently was known only by its catalog designation, 2MASS J23062928–0502285. Even in the images obtained with our largest telescopes and most advanced cameras, the star appears as a nondescript red dot (plate 1). At present, there is no hope of making an image focused tightly enough to allow us to see the planets circling around the star. And yet, using the methods described in this book, we can be certain that this star—now known as TRAPPIST-1—has at least seven planets. We know each planet's mass, diameter, and distance from the star, to within a few percentage points. Two of the planets are orbiting within the star's *habitable zone*, defined as the range of distances where the heat from the star would cause an Earth-like planet's surface temperature to be between about 0° and 100° Celsius, thereby allowing water to exist as a liquid. The word *habitable* is used to describe this situation because liquid water is important for all known forms of life on Earth. This has led many scientists to suggest that in our search for extraterrestrial life, we should prioritize the planets where you could go for a swim.

The quest for extraterrestrial life is usually portrayed in the media as the central activity of exoplanetary science. This is an understandable misconception. While the search for life on other planets is the most enticing goal of exoplanetary science, animating both scientists and the public, the truth is that it remains a speculative endeavor. We've made great strides toward finding Earth-like planets in the habitable zones of Sun-like stars, but honestly, we don't know if finding decisive evidence for *life* on those planets will take decades, centuries, or millennia—or even if extraterrestrial life exists at all. In the meantime, the most remarkable and informative discoveries have been planets with unexpected and un-Earth-like properties. My research group and many others are kept busy studying these strange new worlds, in addition to seeking Earth-like planets. By detecting and studying the full range of possible planetary systems, we hope to learn more about where planets come from, and view the Earth and the Solar System from a more universal perspective.

In writing this book, my goal was to give you a complete briefing on the field of exoplanetary science that is as accurate as possible without requiring specialized training. If I have succeeded, you will be able to understand and enjoy the progress we have made and the breakthroughs you will read about in the future. You'll be able to separate science fact from science fiction and follow the boundary as it keeps moving. You'll understand the scientific principles that allow us to detect and study exoplanets, using instruments ranging from backyard telescopes to billion-dollar spacecraft. You'll see how the new discoveries have revised our understanding of the formation of stars and planets, including our own Solar System. It has been a dream come true to become an astronomer (although I wouldn't have minded becoming an astronaut, either), to participate in the exploration of exoplanets, and to be able to share what we have learned so far.

CHAPTER TWO

THE OLD WORLDS

Before setting out to explore strange new worlds, it will be helpful to re-acquaint ourselves with the planets next door. With a heightened awareness of the properties of the Solar System, and the physical principles that explain those properties, we can appreciate what was considered "normal" and what would have been considered "strange" before the dawn of exoplanetary science. We should keep in mind that four and a half centuries elapsed between the realization that the Earth and the other planets orbit the Sun and the earliest discoveries of exoplanets. During that long interval, generations of astronomers studied the Solar System in ever-greater detail. Generations of mathematicians and physicists refined their theories to explain the phenomena the astronomers were observing, and their theories became deeply entrenched in the scientific culture. To savor the surprising discoveries described later in this book, we need to put ourselves in the mindset of those pre-exoplanet astronomers and theoreticians.

To the naked eye, the planets of the Solar System look just like stars. Both planets and stars appear to be structureless points of light. However, even the earliest stargazers could differentiate between planets and stars, based on the way they move from night to night. The stars are organized into seemingly eternal patterns: the constellations. Ancient astronomers imagined the Earth to be surrounded by an invisible *celestial sphere*, upon which the stars are affixed like Christmas lights. In this scheme, the rotation of the celestial sphere is what causes the constellations to rise in the east and set in the west, as seen from Earth's surface. The planets rise and set, too, but as the days go by, the planets slowly wander through the constellations, like glowworms inching their way around the celestial sphere. This explains the ancient Greek term for planets: *asteres planetai*, translated literally as "wandering stars."

The planets don't wander randomly, though. They always stay within a belt that is wrapped around the celestial sphere's circumference and goes through the constellations of the Zodiac familiar from astrology, such as Aries, Taurus, and Gemini. This belt, known as the *ecliptic*, also contains the Sun, whose location within the constellations can be ascertained at twilight. The planets inch their way around the ecliptic, generally in the same direction as the Sun, but at varying speeds and with occasional reversals in direction. By the early seventeenth century, these observations, along with evidence obtained with the earliest telescopes, persuaded many astronomers that the planets are objects that revolve around the Sun on circular orbits with different speeds—and that the Earth is a planet, too, with its own orbit around the Sun. The celestial sphere appears to rotate because the Earth itself is rotating, and the planets are

14

confined to the ecliptic because their orbits are aligned with one another, forming a single plane. In this respect, the planets' orbits are like different grooves on a vinyl record.[1]

Johannes Kepler, court mathematician of the Holy Roman Empire, enhanced the accuracy of this description of the Solar System in two books published in 1609 and 1619. Using the best available data, Kepler uncovered three regularities in the motion of the planets, which are now called *Kepler's laws of planetary motion*.

Kepler's first law states that each planet traces out an *ellipse*, not a circle, as it goes around the Sun. Viewed from above the Solar System, the orbits of the planets would look circular at first glance, but a careful inspection would reveal that they are slightly squashed (figure 2.1). The diameter of Mars's orbit, for example, is 0.4% longer along its maximum dimension than it is along its minimum dimension. More noticeable is that the Sun is not centered within Mars's orbit; the distance between Mars and the Sun changes by about 20% over the course of a full orbit. To varying degrees, all the planets share these attributes. They follow nearly-but-not-quite circular orbits, with the Sun located off-center.

As far back as the fourth century BCE, ancient Greek philosophers advocated theories in which the planets follow orbits of different sizes, and they were aware of ellipses as mathematical constructions, but I think they would have been surprised to learn that planetary orbits are elliptical. Circular orbits seemed much more natural. Mathematically, a circle is formed by connecting all the points that are the same distance

1. I realize this analogy only works if you're of a certain age. If not, try replacing "vinyl record" with "compact disc," and if that doesn't work, you're on your own. The advent of streaming music has been beneficial to music enthusiasts and detrimental to Solar System analogists.

CHAPTER TWO

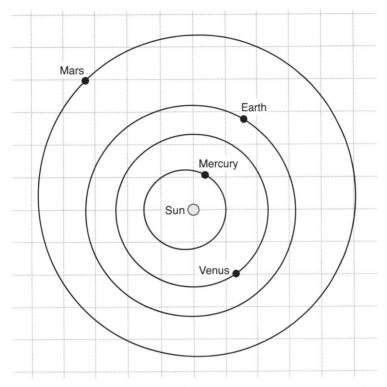

FIGURE 2.1. The orbits of the inner planets are nearly circular, nearly centered on the Sun, and nearly aligned within a single plane. The outer planets obey the same patterns. The orbits are drawn to scale, although not the sizes of the planets or the Sun.

from a chosen center. Imagine sticking a pushpin into a piece of paper on a corkboard. Put a loop of string around the push-pin, pull the string taut with the tip of a pen, and circulate the pen around the paper, drawing as you go. You've made a circle (figure 2.2). The circle's *radius* is the distance between the center and any point on the circle.

THE OLD WORLDS

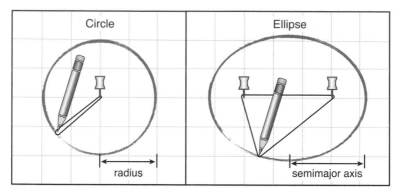

FIGURE 2.2. How to draw a circle or an ellipse using a pencil, two push-pins, and a piece of string. The size of a circle is specified by its radius. The size of an ellipse is specified by its semimajor axis, defined as one-half of the length of the long axis.

To draw an ellipse, start by sticking *two* pushpins into the paper. Loop the string around both pushpins and pull the string taut with the pen, making a triangle. Keeping the string taut, sweep the pen around both pushpins while you draw. You've made an ellipse, centered on the point halfway between the pushpins. With this picture in mind, I can define some terms that will be important for describing planetary orbits. The pushpins are the ellipse's *focal points*. The *major axis* is the ellipse's longest diameter, running through the pushpins from one end of the ellipse to the other end, and the *semimajor axis* is one-half of the major axis. To quantify the size of an ellipse, we usually quote the length of its semimajor axis, just as we use the radius to quantify the size of a circle.

We also need a number to specify the ellipse's shape—whether it is nearly circular or highly elongated. To draw a more elongated ellipse, you need to start with the focal points farther apart. The distance between focal points divided by the length of the major axis is called the *eccentricity*, a number that can be anywhere between zero and one. When the focal points are closely spaced, the eccentricity is close to zero and the ellipse is nearly circular. Moving the focal points farther apart leads to a larger value of the eccentricity and a more squashed ellipse.

Kepler's first law states that each planet in the Solar System follows an elliptical orbit, with the Sun located not at the center, but instead, at one of the two focal points. All the planets have low eccentricities, ranging from 0.007 for Venus to 0.206 for Mercury. The Earth's eccentricity is 0.017, and the length of its semimajor axis is about 150 million kilometers, which by a long-standing tradition is called the Astronomical Unit and abbreviated AU.[2] The AU is the most convenient unit for expressing the sizes of orbits. For example, the lengths of Mercury's and Neptune's semimajor axes are 0.39 AU and 30 AU, respectively. The numbers 0.39 and 30 are much easier to grasp and more helpful for imagining the Solar System than the orbital distances expressed in hundreds of millions of kilometers.

Besides the eccentricity and the length of the semimajor axis, another number associated with each planet's orbit is the *inclination*, the angle between its orbital plane and a reference plane, usually taken to be the Earth's orbital plane. With this definition, the Earth's orbit has an inclination of 0°. All the other planetary orbits have small inclinations, ranging from

---

2. Formally, the AU is defined to be 149,597,870,700 meters, which is approximately the Earth's average distance from the Sun, not the length of its semimajor axis. The distinction isn't important in this book because the Earth's eccentricity is so low.

0.8° for Uranus to 7° for Mercury. This is another way of saying that the Solar System is nearly flat, with all the planets
going around the Sun in the same direction. The Sun rotates
around its own axis in the same direction, too—its equatorial
plane is tipped by only 7° relative to Earth's orbital plane.

To Kepler's predecessors, ellipses would have seemed needlessly complicated. Tradition and intuition wedded them to
the concept of concentric circles. Even when the planetary
data became good enough to determine that the orbits cannot
be concentric circles, theoreticians couldn't bring themselves
to give up on the idea. They doubled down on circles, creating
a model in which each planet moves in a circle, the center of
which is itself moving in a circle. These "epicycles," along
with other embellishments on the idea of circular motion,
became the basis for the Ptolemaic Model for the Solar System that prevailed until Kepler's discoveries.

Kepler's second law specifies how a planet's speed changes
as it moves around the Sun. When a planet is close to the Sun,
it moves quickly, and when it is far from the Sun, it moves
slowly, in a mathematically interesting way. Imagine looking
down on a planet's elliptical orbit from above the Solar System. Draw a line connecting the planet and the Sun. As the
planet revolves around the Sun, the connecting line circulates
around like the minute hand of a clock. Kepler's second law
says that the connecting line "sweeps out area" at a constant
rate. This peculiar phrase means that if the connecting line
were slathered with paint that colors the space it traverses, the
area of the painted region would increase steadily with time.
When a planet is close to the Sun, the connecting line is short,
which by itself would reduce the area-sweeping rate—but the
planet also moves *faster* when it is closer to the Sun, by just
enough to keep the rate constant (figure 2.3).

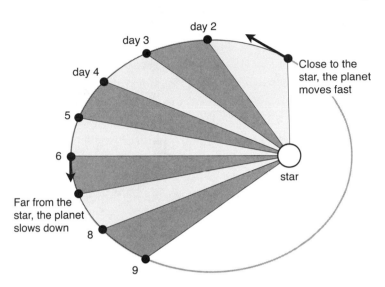

FIGURE 2.3. Kepler's second law states that the line between a planet and the Sun sweeps out area at a constant rate. In this diagram of a fictitious planet, each gray sector is the area swept out over one day. All the sectors have the same area.

This turns out to be equivalent to saying that the connecting line circulates at a rate that varies inversely as the square of the distance from the Sun. For example, the distance from the Earth to the Sun is 0.983 AU in January, and 1.017 AU in July. Kepler's second law tells us that the "minute hand" of the Earth's orbit advances faster in January than in July by a factor of $\left(\dfrac{1.017}{0.983}\right)^2$, which is approximately 1.07. That's why the official lengths of the seasons, defined according to the solstices and equinoxes, are not the same. In the northern hemisphere, winter lasts about 89 days and summer lasts about 94 days, because the Earth moves a little faster in winter than summer.

THE OLD WORLDS

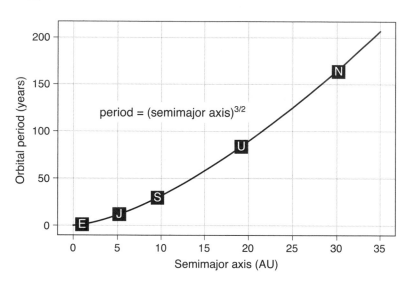

FIGURE 2.4. Kepler's third law states that a planet's orbital period (in years) equals its semimajor axis (in AU) raised to the 3/2 power. The black squares show data for the Earth, Jupiter, Saturn, Uranus, and Neptune.

Kepler's third law is related to the time required for a planet to go all the way around the Sun, which is known as the planet's *orbital period*. Planets on wider orbits have longer orbital periods, in another mathematically interesting way (figure 2.4):

period in years = (semimajor axis in AU)$^{3/2}$

For example, Jupiter's orbit has a semimajor axis of length 5.2 AU. To calculate Jupiter's orbital period, we raise 5.2 to the three-halves power, or, equivalently, we multiply 5.2 by the square root of 5.2. Your calculator can verify that the answer is 11.9, and indeed, Jupiter is observed to take 11.9 years to travel a complete orbit around the Sun.

Proving Kepler's Laws

Where do Kepler's laws come from? Why do planets obediently adhere to these mathematical relationships? It wouldn't have helped to ask Kepler. Although he deserves to be immortalized for uncovering the hidden mathematical patterns in the motion of the planets, his proposed explanations for the patterns—which involved the principles of musical harmony and the proportions of geometrical shapes known as Platonic solids—have not endured. It fell to Isaac Newton, a few generations later, to explain Kepler's laws in terms of more fundamental physical principles. Newton showed that Kepler's laws of planetary motion are consequences of the universal laws of motion and gravitation that Newton promulgated in his masterpiece, *Philosophiœ Naturalis Principia Mathematica*, first published in 1687.

According to Newton's laws of motion, when an object is left undisturbed by external forces, it sits still or moves in a straight line at a constant speed. When an object feels a force—whether it is tugged by a rope, knocked by a collision, or attracted by gravity—the object accelerates in the direction of the force. The magnitude of the acceleration is $\frac{F}{m}$, the force F divided by the object's mass m. Because of the division by m, a given force can accelerate a lightweight object more than a heavy one. That's why your throwing arm can accelerate a baseball to a much higher speed than it can for a bowling ball. Newton also realized that forces always come in equal and opposite pairs. Whenever one object pushes on a second object, the second object pushes back with the same force. The relevant catchphrase is: "For every action, there is an equal and opposite reaction." This law explains the recoil of a gun, the thrust of a rocket, and the fact that you might fall over backward if you actually tried to throw a bowling ball.

An important corollary of Newton's laws is the *conservation of momentum*. Momentum is defined as *mv*, mass times velocity. When several objects interact exclusively with each other in the absence of external forces, their total momentum in any direction remains constant—after taking into consideration that two momenta in opposite directions can cancel each other out. If you were to throw a bowling ball while standing on slippery ice (so there is hardly any friction), the recoil would cause you to slide backward with a momentum that is equal in magnitude and opposite in direction of the ball's forward momentum.

Another corollary of Newton's laws—not as well known, but important for understanding planets and exoplanets—is the conservation of *angular* momentum. Just as an object's momentum only changes when you apply a force, its angular momentum only changes when you apply a *torque*, a turning force. The object could be rotating around an internal axis, like a spinning ice skater, or it could be revolving around an external axis, like a planet going around the Sun. A planet's orbital angular momentum is $mr^2\omega$, where m is the planet's mass, r is the distance to the Sun, and ω is the angular velocity (the rate at which the planet's "minute hand" circulates). For an ice skater, there's a subtlety: there is no single value of r. Different parts of the skater's body are at different distances from the rotation axis. We can still use the same formula, $mr^2\omega$, if we allow r to represent a suitable *average* distance between each part of the object and the rotation axis.[3] For example, if you're twirling on ice skates and pull in your arms, you reduce your body's average value of r. In order to conserve angular momentum, your angular velocity ω must increase in response. You twirl faster.

Newton's laws of motion apply universally, regardless of the origin of the forces and torques.[4] For planetary motion, the relevant force is gravity, the subject of the other breakthrough announced in the *Principia*. Newton's law of gravity states that an object of mass M attracts an object of mass m with a force proportional to $\dfrac{Mm}{r^2}$. According to this equation, the gravitational force weakens as the square of the distance between the masses. If two masses are moved twice as far apart, the attractive force between them is reduced by a factor of four.

The easiest of Kepler's laws of planetary motion for Newton to explain was the second law, about the changing speed of a planet throughout its orbit. Kepler's second law turns out to be a direct consequence of the conservation of angular momentum (although Newton himself did not use that term). Because the gravitational force is always directed toward the Sun, the force points straight at the axis of revolution, without any component that would accelerate the planet's "minute hand" in the clockwise or counterclockwise direction. Therefore, there is no turning component to the gravitational

THE OLD WORLDS

force; it produces no torque. Without any torque, a planet's angular momentum stays the same throughout its orbit. So, if the planet's distance from the Sun decreases, then its angular velocity must increase, just as pulling your arms inward causes you to twirl faster on ice skates. To prevent $mr^2\omega$ from changing, the angular velocity ω must vary inversely with r^2, which is equivalent to Kepler's second law. Voila!

Kepler's first law of planetary motion, the one about elliptical orbits, was much harder for Newton to deduce from the universal laws of motion and gravity. The intense mathematical challenge of this task motivated Newton to invent the subject we now call calculus. The proof is beyond the scope of this book, because I don't want to assume you know (or remember) calculus, and because I'm not aware of a satisfying and concise proof of Kepler's first law that avoids calculus.[5] When I teach my students how to derive Kepler's first law, I spend a half-hour filling the blackboard, and it feels like a miracle when the formula for an ellipse emerges from a thicket of equations. Just before the end, there's a moment when the r^2 from the equation for angular momentum cancels out the $\dfrac{1}{r^2}$ from the law of gravity that always gives me the chills.

Proving Kepler's third law, the one relating orbital period and semimajor axis, is not straightforward either—but at least it can be understood without calculus. There are two reasons why the orbital period (P) grows when the semimajor axis (a) gets longer. First, the orbit has a larger circumference that takes more time to traverse, an effect that by itself would cause P to

---

5. For a satisfying (but lengthy) proof, see the sources listed in Further Reading for this chapter. (For the sake of brevity, I won't keep calling out Further Reading in the footnotes; you might want to check it out after each chapter.)

CHAPTER TWO

be proportional to a. Second, a more distant planet moves more slowly because the force of gravity is weaker at larger distances. The slow-down causes an additional lengthening of P that turns out to be proportional to the square root of a. Multiplying a by the square root of a leads to the $a^{3/2}$ that appears in Kepler's third law.

After Newton's magnificent accomplishments, the planetary orbits posed a new puzzle. Instead of "why are the orbits elliptical?" the question became "why are they so close to being circular?" Nothing about Newton's laws mandates nearly circular orbits. Likewise, nothing in Newton's laws requires all the orbits to lie in nearly the same plane. The Solar System appears to be more orderly and symmetric than required by the universal laws of physics. This curious fact, as we will see, was the main inspiration for the theory of planet formation.

Planets as Worlds

Before getting to the theory, though, we need to discuss another striking pattern in the Solar System that was impossible for Kepler to discover, because his telescope wasn't good enough. As the centuries progressed and technology advanced, planets became more than wandering points of light in the night sky. Viewed through ever larger and more powerful telescopes, and ultimately visited by spacecraft, the other planets became worlds unto themselves.

In elementary school, you probably memorized the names of the planets in order of distance from the Sun: Mercury, Venus, Earth, Mars, Jupiter, Saturn, Uranus, and Neptune (along with Pluto, if you attended school prior to 2005). Many silly mnemonic phrases are available to help with this task, such

as My Very Excellent Mother Just Served Us Nachos. Your
teacher might have asked you to choose your favorite planet
and make a poster about it. What makes this perennial activity
so engaging is that each planet has unique and peculiar characteristics. Mercury is an airless ball of rock and metal that zips
around the Sun every 88 days. Venus's surface is hotter than an
oven, and the rain falling from its thick clouds is composed of
sulfuric acid. Mars has red desert landscapes, blue sunsets, and
a volcano the size of France. Jupiter, a striped and stormy behemoth, is almost all gas, with no surface. Saturn is blessed
with gorgeous rings—that's the planet I chose for my poster,
and it remains my favorite. Uranus is tipped over on its side
for unknown reasons. Neptune is blue, making it appear to be
covered by oceans, although the blue hue is actually caused
by methane gas.

Maybe your class made a scale model of the Solar System,
or maybe you saw such a model at a science museum, in which
the sizes of the Sun and the planets, and the distances between
them, are all in the correct proportions. The point of this
exercise is that the distances between the planets are vast in
comparison with the sizes of the planets themselves. For example, imagine scaling down the Solar System by a factor of
13 billion, a number chosen to make the Earth one millimeter
across, about the size of a poppy seed. Take a moment to ponder the entire Earth—all the cities, continents, oceans, everything—shrunk to fit in the crease of your palm. On the same
scale, the Sun is about the size of a grapefruit and is separated
from Earth by about ten paces. Saturn is a blueberry at a distance of about one football field, and if you kept walking for
another two and a half football fields, you'd reach a pepper-
corn, representing Neptune.

CHAPTER TWO

Scale models involving everyday objects and settings are fun and helpful, and I will refer to them throughout this book. Sometimes, though, one tires of hearing about fruit and football fields and simply wants the numbers. Just as we express the sizes of orbits in terms of AU, a unit based on the size of the Earth's orbit, it will be convenient to express the sizes of planets as multiples of the size of the Earth. Jupiter, for example, is 11.2 times larger than the Earth, so we say that Jupiter's radius is 11.2 Earth radii, or 11.2 $R_\oplus$. The encircled plus-sign, $\oplus$, is the astronomical symbol for the Earth. The Sun's symbol, $\odot$, is also widely used; for example, we write 1 $R_\odot = 109\ R_\oplus$ to say that the Sun's radius is 109 times larger than Earth's radius. Some useful approximations to keep in mind are that the Sun is about 10 times bigger than Jupiter, and Jupiter is about 10 times bigger than the Earth. It's also helpful to know that 1 AU is approximately 200 $R_\odot$, which is to say, the radius of Earth's orbit is approximately equal to 200 times the Sun's radius.

With this scale model of the Solar System in mind, the striking pattern I alluded to earlier is that the four inner planets are small and solid, and the four outer planets are large and gassy.[6] Mercury, Venus, Earth, and Mars have sizes of 0.38, 0.95, 1, and 0.53 $R_\oplus$, respectively (plate 2), and they all orbit within 2 AU of the Sun. They're called the *terrestrial* planets. In contrast, Jupiter, Saturn, Uranus, and Neptune are *giant* planets, with sizes of 11.2, 9.5, 4.0, and 3.9 $R_\oplus$, and orbital distances between 5 and 30 AU. The terrestrial planets are composed

---

6. Our old friend Pluto, a solid object at 40 AU, doesn't fit this pattern. It also has a higher eccentricity and inclination than the planets. These were early clues that Pluto belongs to a different category of objects, the Kuiper Belt Objects, discussed in chapter 8.

THE OLD WORLDS

almost entirely of rock and metal, while the giant planets are bulked up by hydrogen and helium gas. Jupiter and Saturn are almost entirely made of hydrogen and helium, which is why they are called *gas giants*. Neptune and Uranus probably consist of 10–20% hydrogen and helium by mass, with the rest being an unknown mixture of rock, metal, water, methane, and ammonia. Neptune and Uranus are sometimes called *ice giants* because their atmospheres are cold enough for water to freeze—although almost all of the water is buried deep within their interiors, where the high pressure creates exotic phases of matter, unlike ordinary liquid water or solid ice.

The observed connections between planet size, orbital distance, and composition do not follow directly from the fundamental laws of physics. A ball of rock and metal with the mass of Saturn or Jupiter is physically possible, but no such object exists in the Solar System. If a gas giant like Jupiter were placed in a small orbit around a star, or if an Earth-like planet were placed in a very distant orbit, no law of physics would be violated and nothing terrible would happen. That's why the size and compositional ordering in the Solar System intrigued theoreticians. It seemed to demand a special explanation, which, if it could be found, would be a clue about the Solar System's formation.

Formation of the Solar System

The Sun did not come into existence immediately after the Big Bang that began the universe. The Sun and other stars needed to be born. Stars are composed mainly of hydrogen and helium, the only two elements that existed in large quantities in the early universe, and the most common elements in today's universe. When we look around our galaxy, in addition to stars we find patchy clouds of hydrogen and helium gas, hundreds of light-years across. Star formation begins when gravity causes one of those clouds to contract and collapse under its own weight.

Gravity is a universal attractive force. Everything attracts everything else. By itself, this would seem to require everything to draw ever-closer together and merge into a single point, a black hole. There are many ways to avoid this fate, though, because the constituents of the universe are in constant motion, and because gravity is not the only force in the universe. For most of the gas clouds we observe, gravitational collapse is postponed indefinitely, because the gas molecules are moving fast enough and are spread far enough apart for the effects of gravity to be slight and slow to act. It also helps that most clouds are permeated by magnetic fields that resist the compressive force of gravity.

When one part of a cloud happens to become too dense—due to a random fluctuation, a collision with another cloud, or a shock wave from a nearby supernova explosion—then gravity begins to dominate the dense region. The overly strong gravity from the overly massive region starts pulling the rest of the cloud inward. Material from further away crashes into the increasingly dense center, producing shock waves and heat. The cloud breaks into fragments, each of which resumes collapsing. Eventually, the material at the center of each contracting fragment becomes dense and hot enough to ignite nuclear fusion reactions. Hydrogen fuses into helium, liberating lots of energy and sustaining the high temperature and high pressure of the gas. The high pressure allows the hot, dense concentration of mass—a newborn star—to resist further contraction for billions of years.

THE OLD WORLDS

Is that how planets are born, too? Many astronomers used to think that planet formation is a minor-league version of star formation, beginning with the gravitational collapse of a small portion of a cloud of gas that is itself orbiting a newborn star. Variants of this theory have been developed over the last century, and it remains viable as a possible explanation for some of the known exoplanets with the largest masses and widest orbits. The modern consensus, though, is that planets rarely form through gravitational collapse. One reason for this consensus is that the expected temperature, pressure, and orbital speeds of the gas surrounding a newborn star are too high to permit gravitational collapse. In addition, the gravitational-collapse theory predicts that planets should be composed of the same mixture of elements as the Sun and the gas clouds we see strewn about the galaxy—but they're not. The inner planets are composed of rock and metal, with almost no hydrogen or helium. The outer planets have a lot of hydrogen and helium, like the Sun, but they contain heavier elements such as oxygen, carbon, and nitrogen in much higher proportions than the Sun or interstellar gas clouds.

In the modern theory of planet formation, planets arise from processes that are much more complicated than gravitational collapse, which act over millions of years to gather the material left over after star formation. Starting in the 1980s, astronomers found evidence that newborn stars are surrounded by spinning circular disks, with diameters as large as 100 AU (plate 3). Imagine an enormous whirlpool of gas surrounding a young star, swirling faster near the star and slower in the outskirts. The gas slowly spirals inward until it crashes onto the star's surface.

The disks go by several names. Sometimes they're called *circumstellar* disks to convey that they surround stars. Sometimes they're called *accretion* disks to emphasize that the material is accreting onto the star—that is, falling onto the star and adding to its mass. In this book, I'll call them *protoplanetary* disks, because in the modern theory of planet formation, these disks are where planets are born.

Soon, I'll explain how this theory works. But immediately you can see the theory's potential to explain the geometry of the Solar System. Because protoplanetary disks are observed to be *flat*, any planets that form within them would naturally have orbits that are aligned with the plane of the disk. Because protoplanetary disks are observed to be *circular*, with the gas and other material streaming around the star in nearly perfect circles, any planets that congeal out of this material would naturally have circular orbits. Protoplanetary disks are only found around stars younger than a few million years. Beyond that age, almost all the material has either been fashioned into planets, accreted onto the star, or blown out into space by the star's radiation or by the radiation from other stars in the neighborhood. The Sun is 4.6 billion years old, and its protoplanetary disk is long gone, but the circular and neatly aligned orbits of the planets are still intact, testifying that the disk once existed.

Why do protoplanetary disks exist? Why does a gravitationally collapsing gas cloud form a whirlpool, instead of falling directly toward the center? It's for the same reason that the water in your bathtub forms a swirling vortex around the drain and spills in slowly, instead of funneling straight down the drain. In both cases, the key physical principle is the conservation of angular momentum. A galactic gas cloud, like the water in the tub, is never completely at rest. The gas is turbulent, with different parts of the cloud traveling around the galaxy at different speeds. The sum of all these motions is never

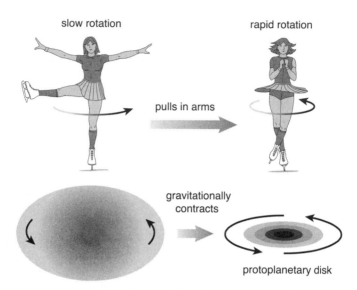

FIGURE 2.5. When ice skaters pull in their arms or gas clouds contract under gravity, they spin faster, because of the conservation of angular momentum. The cloud flattens into a disk because of collisions between streams of gas moving upward and downward. Thankfully, there is no analogous process that flattens ice skaters.

exactly zero. The cloud always has an overall sense of rotation, even if it's very slight. As a result, the cloud has a nonzero total angular momentum.

Because of the conservation of angular momentum, when gravity causes the cloud to contract to a smaller size, the rotation rate is amplified. It's a twirling ice skater on a galactic scale (figure 2.5). In addition, collisions between different parts of the cloud tend to sap away the turbulent motion—for every blob of gas moving one way, there's another blob going the opposite way, and as everything collides and mixes, these relative motions cancel out. Eventually, all the internal motion within the cloud dies down except for the overall sense of rotation, an imbalance that cannot be cancelled completely. The result is a flat, circular, spinning disk.

For this gaseous vortex to drain onto the star, it must shed its angular momentum. In your bathtub, the water loses some of its angular momentum due to friction. The water rubs against the tub's surface, gradually transferring its angular momentum to the tub, and ultimately to the floor to which the tub is affixed, where it makes a negligible impact. Friction also plays a role in a protoplanetary disk. Because the orbital speed of the gas is faster near the star, each ring of gas orbits faster than the ring of gas slightly farther from the star. The relative motion between adjacent rings of gas leads to friction. The inner, faster-moving ring exerts a dragging force on the outer, slower-moving ring, and vice versa. The net effect is to slow down the inner ring and speed up the outer ring, thereby transferring angular momentum outward. Gradually, the disk's angular momentum flows away from the star, allowing the material in the inner part of the disk to fall onto the star. The angular momentum is banished to the disk's wispy outer reaches, where it has a negligible impact.

The shedding of angular momentum is a relatively slow process, because the gas within the disk is tenuous and the frictional forces are weak. As a result, there was a prolonged period lasting several million years when the newborn Sun was still surrounded by a protoplanetary disk. The planets' nearly circular and well-aligned orbits, and the Sun's rotation in the same direction, can all be traced back to the net angular momentum of a particular fragment of the primordial gas cloud that underwent gravitational collapse.

It's a neat theory—but we're not done with it yet. I need to explain how the protoplanetary disk spawned the planets.

How to Make a Planet

Protoplanetary disks contain plenty of raw material that can be used to make planets. Along with hydrogen and helium gas, which account for roughly 99% of the mass, a protoplanetary disk is sprinkled with other elements. The most common elements besides hydrogen and helium are oxygen, carbon, and nitrogen, which often pair up with hydrogen to make the common molecules water (H_2O), methane (CH_4), and ammonia (NH_3). Close to the star, where it's hot, these substances exist as gaseous vapors. Farther away, they freeze into microscopic flakes of ice. There are also atoms of silicon, magnesium, iron, and other elements that combine with oxygen and other atoms to make silicate minerals and metal oxides. These solid materials exist as microscopic grains of dust, which become coated with ice in the colder regions of the disk.

How does Mother Nature start with grains of dust and end up with planets? It's far from obvious how such a feat is achieved. I can't explain it by simply citing a basic principle of physics such as the conservation of angular momentum. The modern theory of planet formation is a lengthy saga involving many principles of physics, chemistry, and material science.

The saga begins with the dust grains orbiting within the gaseous protoplanetary disk. Because the dust grains and their icy coatings are made of heavier elements than hydrogen and helium, they have a higher density than the gas. So, over time, gravity causes the dust to settle down into a more concentrated layer, like the white flecks in a snow globe settling to the bottom. The result is a thinner disk of dust at the midplane of the thicker gaseous disk (figure 2.6). Within this thinner layer, the probability of collisions between dust

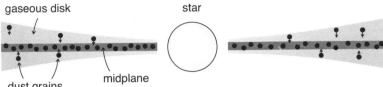

FIGURE 2.6. Side view of a protoplanetary disk around a young star. Within the gaseous disk, the dust grains settle down into a thin layer at the midplane, where they are more likely to collide with each other and stick together.

grains is more likely than when the dust was widely dispersed throughout the gas. What happens when dust grains collide? Collisions at high speed are often destructive, shattering the dust grains into even smaller pieces. However, a solid particle moving through gas feels aerodynamic drag forces, like a bicyclist feeling the effects of air resistance. These drag forces sometimes herd the dust grains together and allow them to collide and merge at low speeds, like bicycle racers clumping together to form a peloton. Through this process, the motes of dust and ice gradually grow into pebbles, rocks, and boulders.

To be honest, this part of the story is a little hazy. We cannot observe this dust-to-boulders process directly. The best we can do is to study meteorites that contain dust grains and pebbles that are thought to have grown within the protoplanetary disk. To help us figure out how they stuck together and grew, we need to rely on computer simulations of solid particles flowing through a gaseous disk—and in such simulations, the small particles do not inevitably merge to form boulders. Often, the particles don't end up clumping together, or if they do, they end up falling onto the star before they can grow very large. The outcome depends on the conditions within the disk and the aerodynamic properties of the solid particles, neither of which is well known.

The story becomes clearer if the solid bodies manage to reach a size of about a kilometer, at which point they're called *planetesimals*. These bodies are massive enough for their own gravity to be important. The gravity of a planetesimal is strong enough to attract any pebbles, rocks, or other planetesimals in the vicinity, causing them to collide even if they were not initially on a collision course. The enhanced collision rate leads to more rapid growth of planetesimals into mountain-sized objects and, ultimately, into objects as large as the Moon or Mars. These *protoplanets* remain after the gaseous disk is gone, and they continue colliding over the next hundred million years. Collisions between protoplanets produce so much heat that the protoplanets melt, at least partially, allowing them to merge and congeal into larger bodies. What started out as microscopic seeds of dust and ice have blossomed into objects comparable in mass to the Earth.

The next chapter in the story is about the formation of the giant planets. This chapter attempts to explain not only where the giant planets came from, but also why they are mainly gaseous, and why they are found exclusively in the outer Solar System. The technical name for this part of the theory is *core-nucleated gas accretion* or simply *core accretion*. The action takes place within the earliest few million years, after some protoplanets have formed, but before the gaseous protoplanetary disk has dispersed. The crucial moment is when a growing protoplanet reaches a mass several times larger than that of the Earth, at which point its own gravity is strong enough to begin attracting the surrounding hydrogen and helium gas. The planet acquires a diffuse halo of gas. If the planet's mass surpasses a critical value, calculated to be about 8 $M_\oplus$ (eight times the Earth's mass), then this process accelerates dramatically. Such a massive planet has a strong enough gravitational field to compress itself to a higher density by pulling the diffuse halo of gas inward to form a denser sheath. When the planet's density increases, its gravitational attraction grows even stronger, causing it to attract even more gas, which compresses and increases the planet's density still further, and so forth. Gas accretion becomes a runaway, out-of-control process that causes the planet to swell up quickly and become a giant planet. The word "quickly," like most modifiers in astronomy, needs to be contextualized; we're talking about a process that might take ten thousand years.

Because of the overwhelming abundance of hydrogen and helium in the protoplanetary disk, runaway gas accretion can increase a planet's mass by as much as a factor of a hundred. And, according to the theory, the only planets with a realistic chance of reaching the critical mass of 8 $M_\oplus$ are those located in the distant parts of the disk, far away from the star. Out there, where it's cold, there's more solid material available. Water, methane, and ammonia are frozen solid and can contribute to the growth of a solid planet. In this sense, the early Solar System, like an alpine mountain, had a "snow line." Beyond it, there was enough snow to pack onto growing planets and make them massive enough to undergo runaway gas accretion.

The calculated position of the snow line in the early Solar System was at an orbital distance of about 3 AU, based on the estimated heating power of the young Sun and the conditions inside the protoplanetary disk. This places the snow line somewhere between the current orbits of Mars and Jupiter—which nicely matches the boundary between the terrestrial planets (Mercury, Venus, Earth, and Mars) and the giant planets (Jupiter, Saturn, Uranus, and Neptune).

The modern theory of planet formation was a collective achievement of many physicists and astronomers over several centuries. It's a thrilling story. The protagonist, a microscopic speck of dust, embarks on an adventure with many twists and turns, ultimately joining forces with countless other specks to become the core of mighty Jupiter. The theory fits the facts. It links together the formation of the Sun, the rocky planets, and the giant planets into a unified narrative.

Nevertheless, we should maintain a healthy skepticism. I've already alluded to one of the hazy and possibly unrealistic parts of the story, in which solid particles clump together and grow instead of shattering each other or falling onto the star. Even more worrying is that the theory was *invented* to fit the facts. Before the mid-1990s, the only available facts were the facts of the Solar System. But at least the theory made clear predictions. The phenomena of gravitational collapse, protoplanetary disk formation, planetesimal growth, and core-nucleated accretion were portrayed as natural and inevitable outcomes of basic physical and chemical processes. The theory did not invoke rare and extraordinary events, and therefore, the theory predicted that the patterns we observe in the Solar System should be universal.

At last, we can put ourselves into the mindset of astronomers in the mid-1990s. We grew up with the Solar System and the theory of planet formation. We know the true test of a theory is whether it makes correct predictions. Will we observe the same patterns in exoplanetary systems that we see in the Solar System?

CHAPTER TWO

CHAPTER THREE

THE AGE OF EXPLORATION

The year 1995, like 1492, marked the beginning of an age of exploration. The new explorers, instead of using seagoing vessels to reach distant continents, used telescopes to discover planets revolving around distant stars. Two of the earliest exoplanet explorers, Michel Mayor and Didier Queloz, were awarded half of the 2019 Nobel Prize in Physics for the landmark discovery they announced in 1995. My colleagues and I are united in our admiration for their pioneering work, and in our pride to be continuing what they began. However, there's something peculiar about their Nobel Prize citation. It says, "for the discovery of an exoplanet orbiting a solar-type star." Why doesn't it say, "the *first* discovery"? After all, it's not such a big deal anymore to have discovered *an* exoplanet. We know of thousands of exoplanets. New ones are reported every week. These days, an amateur astronomer or enterprising high school student can discover an exoplanet by sifting through publicly available databases.[1] Did the Nobel Committee accidentally leave out the word "first" in the citation?

I don't know for sure. They didn't ask for my input. I suspect, though, that the omission was deliberate—and thereby hangs a tale.

Just as it is difficult to say who discovered America (Christopher Columbus? John Cabot? Leif Erikson? Amerigo Vespucci, whose name is the one that stuck? The people who came on foot from Siberia?), it is difficult to say who was the first person to discover an exoplanet. There are many credible contenders. It's fun to consider the merits of each case, and it provides an opportunity to learn about the methods astronomers use to detect exoplanets. So, let's pretend we are on the Nobel Committee, charged with picking the winner.

The Direct-Imaging Method

First, we need to understand why detecting exoplanets was difficult enough to merit consideration for the most prestigious prize in science. How would *you* try to detect an exoplanet? The first approach that might come to mind is to gain access to one of the world's best telescopes, use it to take pictures of nearby stars, and then search the pictures for planets. The star would appear as a bright dot in the center of an image, and if you looked closely enough, maybe you would be able to see fainter dots near the star. You might have already had such an experience if you've ever viewed Jupiter through a telescope.

---

1. If you'd like to try your hand at detecting exoplanets, a good place to start is the Planet Hunters website (planethunters.org).

(If not, get thee to a telescope; you are missing out on one of life's treats.) You can see Jupiter, the striped behemoth, along with as many as four fainter white dots, just as Galileo Galilei saw for the first time in 1610. The fainter dots are Jupiter's largest moons, Io, Europa, Ganymede, and Callisto, which are called the *Galilean moons*. Their orbits are nearly circular and are aligned with each other, as in the Solar System. From Earth, we see their orbits from the side. A straight line drawn from Earth to Jupiter (our *line of sight*) skims across the moons' orbital planes. That's why the moons appear to march back and forth along a nearly straight line from night to night, sometimes crossing in front of Jupiter and other times disappearing behind it.

We would love to have such direct views of exoplanetary systems. Unfortunately, in practice, this seemingly straightforward approach to finding exoplanets—known as the *direct-imaging* method—is extraordinarily difficult. Planets are not just "fainter" than stars. They are much, much fainter. Jupiter is 1,500 times brighter than Io, the innermost Galilean moon. To an extraterrestrial version of Galileo, looking toward the Solar System from afar, the Sun would be 200 *million* times brighter than Jupiter, and five *billion* times brighter than the Earth. Neither our eyes nor our best telescopes would be able to detect such faint dots—they would be lost in the Sun's dazzling light. Remember the analogy of the firefly and the searchlight?

The fundamental problem is that planets are tiny and cold, compared to stars. When we see one of the planets of the Solar System in the night sky, we're looking at reflected sunlight, rather than the planet's own glow. Because the planets are so small compared to the sizes of their orbits, they intercept and reflect only a teensy fraction of their star's total light output. The same is true of exoplanets.

You might think we could solve that problem by building bigger telescopes that are more capable of concentrating the light from faint astronomical objects. Alas, a bigger telescope, by itself, does not solve the problem. In addition to building a bigger telescope, we need to attach a camera capable of making images with extremely high contrast, by which I mean that the images must faithfully represent very bright objects and very faint objects at the same time. It might not be obvious why this is so difficult. Why does it take high technology to make high-contrast images?

Ideally, we would focus the light from the star onto a single pixel at the center of the image, leaving all the other pixels dark. That way, any additional light from a planet would stand out somewhere else in the image. For a variety of reasons, though, we cannot focus the light from a star as sharply as we wish. Some of the starlight inevitably spills away from the center of the image and covers the surrounding pixels, overwhelming any light from a planet.

One reason for our lack of control over the starlight is the Earth's atmosphere. When light waves travel through air, they are deflected by an amount that depends on the local temperature, density, and humidity. Those properties of the air are always changing, both from place to place and in time. As a result, even though the light waves from a distant star might arrive at the Earth having traveled in a perfectly straight line for millennia, the waves are scrambled by the Earth's atmosphere in the last millisecond of their journey, just before they enter our telescopes. This scrambling effect causes the image of a star to vary in brightness and dance around on the camera's light-sensitive surface. In more familiar language, stars twinkle.

The twinkling problem can be solved by putting a telescope above the Earth's atmosphere, which helps to explain why astronomers love the Hubble Space Telescope, launched in 1990, and the Webb Space Telescope, launched in 2021. Space telescopes are very expensive, though, and even a space telescope cannot achieve an arbitrarily tight focus. The telescope's mirrors and lenses are never completely still or perfectly smooth. They're designed to be smooth surfaces with the right geometry to redirect the incoming light onto the detector, but any irregularities—dust or scratches on the mirrors, degradation of reflective coatings, or flexing of the mirrors—cause light to go where it doesn't belong. Especially troublesome are certain optical defects that corral the stray light into faint dots ("speckles") that look just like planets.

Even if we could build a defect-free telescope and launch it into space, we would still be stymied by *diffraction*, the ultimate limit imposed by the laws of optics on our ability to focus images. Diffraction refers to the tendency for waves to bend and spread out when they encounter obstacles. When light waves enter the circular opening of a telescope, the circular edge acts as an obstacle, and the light waves are unable to pass through on perfectly straight trajectories. Instead, they splay out, making the light from a star appear to come from a range of different angles in the sky, instead of a singular location. In other words, the image of the star is blurry (figure 3.1). Diffraction is also why stars sometimes look spiky in astronomical images, with rays of light shooting out in different directions. The spikes are often caused by diffraction around the posts that support one of the mirrors inside the telescope.

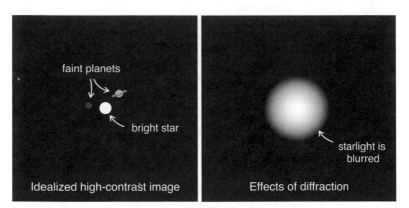

FIGURE 3.1. On the left is an idealized image, in which planets are apparent as faint points of light. On the right is a more realistic image, in which the inevitable phenomenon of diffraction causes the starlight to overwhelm the fainter light from the planets.

Finding planets in the presence of all these blurring effects is extremely difficult, bordering on the impossible, although it is not *totally* impossible. We'll see in chapter 8 that technological advances have led to the detection of a few dozen planets as faint dots in the vicinity of nearby stars. However, the direct-imaging method is not the source of most of our current knowledge about exoplanets, and it was not the method that earned Mayor and Queloz their Nobel Prize. They used a sneakier approach based on Kepler's laws of planetary motion and Newton's laws of motion and gravity.

The Astrometric and Doppler Methods

Kepler's first law states that each planet in the Solar System travels on an elliptical orbit. Newton taught us why: the Sun pulls on the Earth with a gravitational force that weakens with the square of the distance between them. Newton also taught us that for every action, there is an equal and opposite reaction. Together, Newton's laws of gravity and motion imply that something remarkable is happening in the Solar System— something that probably would have blown Kepler's mind. The Sun is moving. If the Sun's gravity is attracting a planet, then the planet's gravity must be attracting the Sun with the same force, causing the Sun to accelerate. All the planets pull on the Sun, causing the Sun to perform a complex dance of looping motions instead of following an elliptical orbit. Because of the Sun's enormous mass, its acceleration is small, and it moves relatively slowly. That's why Kepler was able to fit the available data with his mathematical model despite his mistaken assumption that the Sun sits still.

A more accurate description of the Solar System is that the planets and the Sun all orbit around their collective *center of mass*. The center of mass is the average location of all the bodies of the Solar System, with each body contributing to the average in proportion to its mass.[2] Jupiter, for example, has a mass of 318 $M_\oplus$ and therefore has 318 times as much influence as the Earth in determining the location of the center of mass. If you built a scale model of the Solar System using spheres with the correct relative masses that are glued to a lightweight and rigid disk, the center of mass is where you could balance the disk on your fingertip (figure 3.2). Each planet's elliptical orbit has a focal point located at the center of mass, not the center of the Sun.

2. For example, to find the x-coordinate of the center of mass, multiply the x-coordinate of each body by its mass, add the results, and then divide by the total mass of all the bodies.

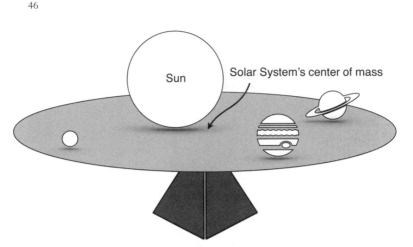

FIGURE 3.2. The center of mass is the average location of all the bodies in a system, with each body contributing to the average in proportion to its mass. If a scale model were placed on a rigid disk of negligible mass, the center of mass would be at the disk's balancing point.

Because the Sun is by far the most massive component of the Solar System, with a mass of 330,000 $M_\oplus$, the Sun is always close to the center of mass. Still, the Sun does shuffle around. It typically moves a distance equal to its own diameter over the course of a few years (figure 3.3). Likewise, any planets around a distant star would cause the star to shuffle around. Detecting the star's motion was the basis of the sneaky method employed by the exoplanet pioneers.

The earliest attempts to discover exoplanets were based on trying to detect the shuffling motion of a star in a series of images. Even if a star's planets are too faint to see, the star should shift its position in the image as it shuffles around the center of mass of its planetary system. This is in addition to the parallax-induced motion described in chapter 1. Trying to detect the planet-induced motion is called the *astrometric*

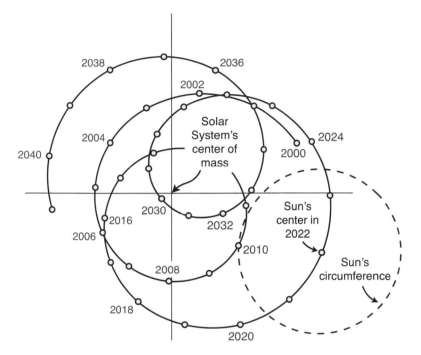

FIGURE 3.3. The Sun, pulled by all the planets, travels on a complicated looping path around the Solar System's center of mass. The distance from the Sun's center to the Solar System's center of mass is never more than a few times the Sun's radius.

method, because the term "astrometry" refers to measuring a star's location in the sky—its latitude and longitude on the celestial sphere. From the mid-nineteenth century through the 1970s, practitioners of the astrometric method announced many exoplanet discoveries, none of which withstood further scrutiny. Measuring the slight displacement of a blob of starlight on an astronomical image turned out to be too difficult and prone to error. One of the most serious problems was, again, the variability of the Earth's atmosphere.

THE AGE OF EXPLORATION

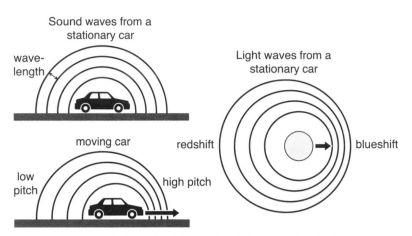

FIGURE 3.4. The Doppler effect is the shift in wavelength that occurs whenever a source of waves is moving toward or away from an observer.

There's another method to detect the motion of a star that doesn't suffer nearly as much from the effects of the Earth's atmosphere. It's based on the *Doppler effect.* Whenever a moving object emits waves—such as sound waves, radio waves, or light waves—the length between successive waves is compressed in the forward direction of motion and stretched in the backward direction (figure 3.4). For sound, the wavelength determines the pitch, explaining why the sound of a car horn or an ambulance's siren falls from high pitch to low pitch as the vehicle approaches and then speeds away. Using radio waves, the same phenomenon allows a police officer to check the speed of a car and a meteorologist to measure the speed of a stormfront.

For light, the wavelength determines the color. When a luminous object moves toward an observer, the light appears ever-so-slightly bluer, and when the object recedes, the light reddens. We never notice this, though, because the fractional change in wavelength is equal to the speed of the moving object divided by the speed of light, and the speed of light is much faster than the speeds we encounter in everyday life. When a car is approaching at 100 km/hour (60 mph), its headlights technically do appear bluer, but the shift in wavelength is only 0.00001%.

Our eyes cannot sense such tiny shifts in wavelength, but astronomers can sense them using specialized equipment. The best current instruments can sense changes on the order of one part in a billion, by which I mean the change in wavelength divided by the original wavelength is on the order of 0.000000001, or, equivalently, 10^{-9}. (The exponent -9 tells us to start with 1 and move the decimal point to the left by 9 places.) Such exquisite precision is only possible because of a marvelous property of starlight. When we use a device called a *spectrograph* to spread out starlight into a spectrum of colors, we see not only a continuous rainbow, but also dark lines or bands, as though a bar code were imprinted on the rainbow (plate 4). The bar code is the star's set of *spectral absorption lines*— highly specific colors that appear to be missing from the rainbow. The lines occur because each substance in a star's outer atmosphere preferentially absorbs and emits light with distinct wavelengths. Hot hydrogen gas, for example, strongly absorbs light with a wavelength of 0.656 μm (millionths of a meter), corresponding to a shade of dark red. Sodium pays little attention to red light, but is fond of a yellow-orange hue with a wavelength of 0.589 μm. Not coincidentally, this is the same color emitted by the sodium lamps that are sometimes used to illuminate highways and tunnels.

Why are chemical elements so fussy about the wavelengths of light they absorb and emit? The answer lies in the domain of *quantum mechanics*, the mathematical theory that governs the behavior of electrons and other fundamental particles. I'll have more to say on this topic in chapter 6. For now, it will suffice to know that spectral absorption lines provide convenient reference markers in the spectrum of a star. They're sharp features that can be precisely tracked over time, to see if they are shifting back and forth in wavelength as the star is pulled around by a planet (plate 5). Together, the Doppler shift and the existence of spectral absorption lines are a powerful combination for exploration, like the billowing sails and sturdy hulls of the galleons that crossed the Atlantic.

The Earliest Known Exoplanets?

In 1980, Gordon Walker, a professor at the University of British Columbia, began searching for planets with his colleague Bruce Campbell and a few other research associates. They were the world's preeminent experts in precise Doppler observations of stars. Walker designed his survey on the premise that all planetary systems resemble the Solar System. Remember the mindset of pre-exoplanet astronomers: they expected planetary orbits to be nearly circular and to lie in the same plane, with the giant planets (like Jupiter) on the outside, and the terrestrial planets (like Earth) closer to the star. Under those assumptions, the biggest signals—and, given the limitations of his equipment, the only signals Walker had any chance of detecting—would come from giant planets. And giant planets were expected to have long orbital periods. Jupiter takes 12 years to go around the Sun, and the other giant planets take even longer. This meant Walker needed to commit himself to a very long-term project.

CHAPTER THREE

He also had to hope that Jupiter-like planets are common, because he was only able to monitor 23 stars. To expand his search, he would have needed more support from the Telescope Time Allocation Committee, which was not forthcoming. Observing time on major professional telescopes is a scarce resource. Part of being an astronomer is the semiannual ritual of writing proposals to gain access to appropriate telescopes. Another part of being an astronomer is occasional "jury duty"—serving on the Time Allocation Committees that decide which proposals to accept. Like many other collective decision-making bodies, these committees tend to be conservative; they prefer to support projects that seem assured of success. Searching for planets did not fit that description. Reflecting on those lean years, Walker later wrote: "It is quite hard nowadays to realize the atmosphere of skepticism and indifference in the 1980s to proposed searches for planets. Some people felt that such an undertaking was not even a legitimate part of astronomy." The precision Walker was aiming for seemed out of reach, and there had already been many claimed detections of exoplanets that turned out to be false.

Despite these obstacles, by 1987, Walker and his team had identified a few stars whose spectral absorption lines appeared to be shifting back and forth in a manner consistent with Doppler shifts caused by planets. "A team of Canadian astronomers say they have found strong evidence of the existence of planets outside the solar system: signs that celestial bodies 300 to 3,000 times as massive as Earth are orbiting up to seven stars," reported the *New York Times*. The clearest signal was from a star called Gamma Cephei, the third-brightest star in the constellation of Cepheus, which appeared to have a Jupiter-mass planet circling around every 2.7 years.

THE AGE OF EXPLORATION

Should Walker have won the Nobel Prize? Was his team the first to discover an exoplanet? It's complicated. It depends on what we mean by the word *discover*.

A dictionary definition is "to obtain knowledge for the first time." Okay, but what is *knowledge*? For that, we open a philosophy textbook to the chapter on epistemology, where we learn that an early definition of knowledge (dating back to Plato) is "justified true belief." Our task, then, is to identify the first person or group of people who held a justified true belief in the existence of an exoplanet. Because it must be *true*, we need not consider the false starts and mistaken claims that tarnished the reputation of planet hunters. By requiring it to be *justified*, we set aside the lucky guessers. In 1953, Philip K. Dick wrote a story set on a planet around the star Proxima Centauri, and in 2017, astronomers detected such a planet, but we can all agree that Philip K. Dick did not discover the planet.

So, now we can sharpen our question: did Gordon Walker have a justified true belief in the existence of a planet around Gamma Cephei?

His belief was likely to be *true*: the planet's existence has been confirmed by gathering more Doppler data over the subsequent decades. A study published in 2003, based on five times as much data as Walker had in 1987, stated: "This supports the planet hypothesis for the residual radial velocity variations for Gamma Cephei first suggested by Walker et al." The term *radial velocity* means the component of the star's velocity that is directed either toward or away from us, the only type of motion that produces Doppler shifts.

The stumbling block in Walker's claim to the prize is whether his belief was *justified* in 1987. At that time, Walker worried that he was being fooled. Not long after the *New York Times* article, he and his colleagues found that at least five of the seven planet-like signals were probably spurious. The Doppler signal from Gamma Cephei seemed unimpeachably strong, but Walker worried that the signal was from something else besides a planet. For example, maybe the 2.7-year pattern was from the star's rotation, rather than motion induced by a planet. The Sun and similar stars rotate about once per month, which is much shorter than 2.7 years, but Walker thought Gamma Cephei was a giant star, and giant stars tend to rotate very slowly. In a 1992 paper, based on this concern and mindful of the history of false alarms in this field, Walker backed away from the claim that the signal from Gamma Cephei arose from a planet.

As it turns out, the star had been misclassified. It's a big star, but not as big as Walker had thought, making it unlikely that rotation was causing the Doppler signal. Walker had been fooled into worrying he was fooling himself. His caution was admirable, and natural, given the atmosphere of skepticism bordering on hostility. Maybe it is best to say that Walker *detected* the planet but did not quite *discover* it.

That brings us to the next contender, David Latham, of the Smithsonian Astrophysical Observatory. In 1989, he and his collaborators reported an intriguing Doppler signal around a star called HD 114762.[3] The signal was crystal clear. Its characteristics were compatible with orbital motion, and incompatible with rotation. This would seem to be a slam-dunk discovery. Indeed, at the time of the Nobel Prize announcement in 2019, the object orbiting HD 114762 was included in

---

3. I'm afraid that this mind-numbing type of name is typical. There are simply too many stars to assign imaginative names to all of them. Instead, the tradition for naming a star is to identify a star catalog (in this case, the Henry Draper or HD catalog) and give the numerical position of the star within the catalog (in this case, the 114,762nd entry).

NASA's most comprehensive database of confirmed exoplanets and was logged with 1989 as the year of discovery. That's six years before the 1995 discovery cited in the Nobel Prize citation for Mayor and Queloz.

Why, then, did Latham not receive an invitation to shake the hand of the King of Sweden? One problem was that it took a long time for his team's accomplishment to be perceived by many astronomers as a planet discovery. There were some peculiar things about the putative planet. To begin, the orbit is far from circular. It's an ellipse with an eccentricity of 0.3, which is higher than the eccentricities of the planetary orbits in the Solar System. Also, the size of the Doppler shifts indicated that the orbiting object—which I'll call Latham's object, for simplicity—was at least 11 times more massive than Jupiter. A mass of 11 Jupiters seemed outlandishly high for a planet, and that was the *minimum* possible mass, because of an important limitation of the Doppler method. The star's velocity around the center of mass can be in any direction, depending on the orientation of the orbit, but the Doppler effect only arises from the star's *radial* velocity—the component of its velocity directed toward or away from the observer. The Doppler effect is blind to any lateral motion. Because the observed radial velocity is lower than the star's three-dimensional velocity by an unknown factor, the observations only allow us to set a lower limit on the mass of the orbiting body.

Finally, the orbit of Latham's object seemed way too small for a giant planet. With a semimajor axis of 0.4 AU, its orbit is less than one-tenth of the size of Jupiter's orbit. According to the theory of core accretion, discussed in the previous chapter, giant planets shouldn't be able to form so close to a star. The inner part of a planetary system was supposed to be the exclusive domain of smaller terrestrial-type planets.

CHAPTER THREE

Despite all these peculiarities, Latham thought that his object might be a planet. Some of his team members (and many other astronomers) thought that was a stretch. As a result, the 1989 paper only mentions the possibility of a planet by way of speculation. More likely, they wrote, it was a *brown dwarf*, a sort of "failed star" with a mass and interior temperature too low to ignite the nuclear fusion reactions that generate power within the Sun.[4] The title of their paper was: "The unseen companion of HD 114762: a probable brown dwarf."

Today, though, none of the peculiarities noted above are considered too peculiar. The NASA exoplanet catalog contains hundreds of giant planets with smaller or more elliptical orbits than any of the planets of the Solar System, including some planets that are even more massive than 11 Jupiter masses. Nevertheless, even if Latham's object were retrospectively designated a *true* exoplanet, and even though Latham's claim to have discovered it was amply *justified* by his team's high-quality data, the fact remains that the claim was *not believed*, partly because of the prejudice that planets should look and act like the planets in the Solar System.

By the time of the 2019 Nobel Prize announcement, astronomers had gotten over this prejudice, but the prize committee still faced a quandary: the object's true mass remained unknown at the time. That's probably one reason why they chose Mayor and Queloz instead of Latham. Interestingly, though, Mayor and Queloz didn't measure the true mass of their prize-winning planet, either. Their discovery was also based on the Doppler method. So, what was the difference?

---

4. The term "brown dwarf" is a misnomer. Brown dwarfs aren't brown. If I were in charge, I'd rename them *infrared dwarfs* because, being cooler than ordinary stars, they emit mainly infrared radiation instead of visible light.

I'm getting ahead of myself. Before getting to Mayor and Queloz, I'd like to return to the story where we left off, in 1989. What happened next was a stunning surprise.

The Freakiest Planets

In 1992, astronomers Aleksander Wolszczan and Dale Frail announced the discovery of two planets around a star located 2,300 light-years away, which they had been monitoring for two years using a variation on the Doppler method. Even at the time, their data made clear that the orbiting objects have masses typical of planets, rather than brown dwarfs or stars, and further observations have strengthened the conclusion. The stunning thing was that the star is not an ordinary star like the Sun. It's a *neutron star*, named PSR 1257+12.[5]

Neutron stars are among the most exotic entities in the universe. They are remnants of supernova explosions, which take place when a massive star runs out of nuclear fuel and becomes unstable. A neutron star packs the mass of an entire star into a ball only about 20 kilometers across, making it so dense that it's on the verge of being overwhelmed by its own gravity. With one false move, it would collapse and become a black hole. As if that weren't dramatic enough, some neutron stars are observed to spin as fast as hundreds of times per second, and others spew forth X-rays and lethal doses of radiation.

The neutron star's radio waves were what enabled the stunning discovery. Neutron stars don't always emit radio waves,

---

5. In this name, PSR stands for pulsar, and the numbers are the first few digits of the object's coordinates in the sky.

CHAPTER THREE

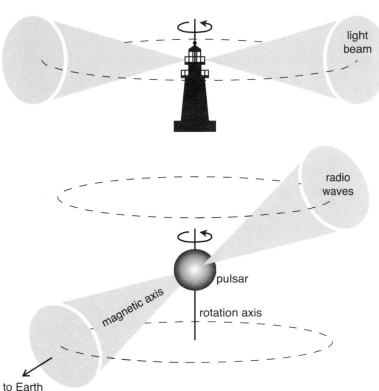

light
beam

radio
waves

pulsar

magnetic axis

rotation axis

to Earth

FIGURE 3.5. Some neutron stars emit radio waves in two rotating lighthouse-like beams. A pulsar is a special case in which one of the beams periodically points in the direction of Earth, allowing us to observe pulses of radio waves.

but when they do, the radio waves are launched into space in narrowly defined directions, pointing north and south along the axis of the star's magnetic field. If the star is spinning around a different axis, then the beams of radio waves swing around like the rotating beam of light from a lighthouse (figure 3.5).

THE AGE OF EXPLORATION

If we are lucky, one of those beams of radio waves periodically sweeps across the Solar System, thereby sending us periodic pulses of radio energy that radio astronomers like Wolszczan and Frail can detect. The members of this rare subclass of neutron stars are called *pulsars*.

If a pulsar were steadily rotating and not otherwise wobbling or accelerating, the pulses would be strictly periodic; the time between any two successive pulses would be the same. An orbiting planet spoils the periodicity by causing the neutron star to accelerate around the system's center of mass. When it is moving toward us, the star is racing to catch up with its own previously emitted pulses. Consequently, the time between successive pulses received on Earth is shorter than average. Likewise, when the neutron star is moving away from us, it races away from its own Earthward-bound pulses, and we observe a longer-than-average time interval between pulses. The pulsar-timing method is conceptually similar to the Doppler method, but instead of observing the periodic rising and falling of the wavelengths of spectral absorption lines, we observe the periodic rising and falling of the pulse rate. Numerically, for PSR 1257+12, the average time between pulses is 6.219 seconds, and the fluctuations due to the planets are about 0.002 seconds in either direction. That may sound like a small effect, but the radio pulses are so sharp that fluctuations can be detected even when they are as small as 0.0001 seconds.

Getting back to our task, how should we score Wolszczan and Frail's discovery as a candidate for the earliest discovery of exoplanets? Their claim was true and justified. It was believed in 1992, and it is still believed by the astronomical community. The only hang-up is whether the objects orbiting the pulsar should qualify as planets.

CHAPTER THREE

Until that point, the astronomical community's working definition of a planet was an object with a mass larger than that of an asteroid and smaller than that of a brown dwarf. The pulsar discovery forced a more careful appraisal. Maybe the word "planet" should be reserved for objects orbiting a normal star, and not a neutron star? (Now you understand why the 2019 Nobel Prize citation refers to "an exoplanet *around a solar-type star.*") Some astronomers insisted that the definition of a planet should include a requirement that it formed within a protoplanetary disk surrounding an ordinary star. Presumably, the objects orbiting the pulsar did not form in this way. They probably formed after the supernova explosion that created the neutron star, because any preexisting planets would not have survived the catastrophic blast of energy. The best current hypothesis is that some of the exploding material ended up falling back down into an orbit around the pulsar, where it formed a second-generation protoplanetary disk that eventually spawned planet-mass objects.

Because there is no universally agreed-upon theory of planet formation, many astronomers (myself included) think it is premature to employ a formation-based definition. What ended up happening, in practice, is that astronomers became comfortable referring to Wolszczan and Frail's objects as *pulsar planets.* But the pulsar planets are considered freaks, and the search for more of them has been frustratingly unproductive, so far. Only a few other pulsars are suspected of having planets, and even in those cases, the evidence is not as secure as it is for PSR 1257+12. Finding more will be difficult because pulsars are only a small minority of neutron stars, and because even among pulsars, few have beams that rotate with the extraordinary regularity that is needed for Wolszczan and Frail's technique to succeed.

A Hot Jupiter

We'll return to the subject of planets around different types of stars in chapter 7. For now, let's resume the story of the exoplanet pioneers, and advance the clock to 1995. Michel Mayor and Didier Queloz, two astronomers at the Geneva Observatory in Switzerland, had been working on improving the Doppler technique. Earlier, Mayor had contributed some data to Latham's study of HD 114762, helping to discover the "probable brown dwarf." Afterward, he and his graduate student, Queloz, decided to try their own luck at hunting for brown dwarfs and maybe even planets. They and their colleagues in Geneva had built a new spectrograph for a telescope at the Haute-Provence Observatory in southeastern France, which improved the precision of their Doppler measurements and created goodwill with the telescope's Time Allocation Committee. This allowed them to monitor more stars than Walker or Latham had done, which turned out to be a crucial advantage.

One of their stars, a Sun-like star named 51 Pegasi (so named because it appears in the constellation Pegasus) was moving to-and-fro with a speed of 60 meters per second and an orbital period of only 4.2 days. These characteristics implied the existence of a planet with a minimum mass in between the masses of Saturn and Jupiter. Such a mass made astronomers comfortable calling the object a planet—more comfortable than they had been with Latham's object. Less comfortable, though, was the short orbital period of 51 Pegasi. According to Kepler's third law, the 4.2-day period implied an orbital distance of only one-twentieth of an AU, or about 100 times smaller than Jupiter's orbit around the

Sun. That, insisted many theorists, was no place for a giant planet. And yet, there it was.

Being so close to the star, the planet orbiting 51 Pegasi is as hot as a pottery kiln. Mayor and Queloz had discovered what would come to be known as a *hot Jupiter*, a type of giant planet that had been presumed impossible according to the core-accretion theory of giant planet formation. As described in chapter 2, this theory held that giant planets should only be able to form beyond a distance of a few AU from a star—beyond the snow line of the protoplanetary disk—where there is plenty of frozen solid material available to pack onto a growing planet and allow it to become massive enough to accrete the surrounding hydrogen and helium gas.

Some astronomers were initially skeptical of Mayor and Queloz's discovery, not only because of the contradiction with their expectations, but also because of the checkered history of the field. One concern was that the Doppler shifts might have been caused by stellar rotation or pulsation, rather than orbital motion. Astronomers gathered more data over the next year, and their concerns were put to rest. The star's rotation rate was found to be too slow to explain the Doppler shifts. Pulsations were ruled out, too. Some stars have outer layers that periodically bulge outward and sink back down, producing Doppler shifts unrelated to orbital motion, but pulsations tend to be somewhat irregular, and the signal coming from 51 Pegasi was as steady as a clock. Pulsations also would have caused subtle changes in the star's spectrum because the bulging parts of the star cool off and the sinking parts heat up. Reassuringly, the spectrum of 51 Pegasi didn't change appreciably, apart from the overall Doppler shifts. 51 Pegasi was the real deal. Planet formation theory would need to be revised. Mayor and Queloz, then, were the first to hold a justified true belief in the existence of an object that nearly everyone agreed and still agrees is an exoplanet orbiting a Sun-like star.[6]

Just as important, the discovery of the planet around 51 Pegasi had the same effect as the first sighting of an unexplored and seemingly boundless continent. The exponential growth rate of planet discoveries and of the number of scientists working in this area began in 1995. Mayor and Queloz led the way, and their group racked up dozens of planet discoveries in the years that followed. That's probably why the Nobel committee thought Mayor and Queloz deserved the scientific spotlight and a half million dollars.

Putting on my pedant's hat, though, I note that Mayor and Queloz's claim that the unseen companion of 51 Pegasi is a planet was not 100% justified. Remember, as was the case with Latham's object, the Doppler method only specified the orbiting body's minimum possible mass. The true mass could have been larger—much larger, if the orbital plane happened to be oriented nearly perpendicular to our line of sight. In that case, the star's motion around the center of mass would be mainly sideways, instead of toward or away from us, causing the Doppler shifts to be small even if the orbiting body were too massive to be a planet. While this would require an unlikely coincidence, it was conceivable at the time that the companion of 51 Pegasi was a brown dwarf.

---

6. Well, you can never convince *everyone*. There is a well-known astronomer who, as recently as 2011, fiercely doubted the existence of hot Jupiters and called them "cryptoplanets."

CHAPTER THREE

In fact, it *is* a planet. In 2017, a group led by Jane Birkby, then at Leiden Observatory, made a decisive measurement of the mass of the planet orbiting 51 Pegasi. Using a very large telescope,[7] they obtained so many spectra that by combining the data, they could see spectral absorption lines from the *planet* in addition to those from the star. This allowed them to monitor the Doppler shifts and measure the radial velocities of both the planet and the star. Although the radial velocity is lower than the true velocity by an unknown factor, the factor is the same for both the star and the planet. This means the ratio of the radial velocities is the same as the ratio of the full three-dimensional velocities—the unknown factor cancels out. Birkby's team found that the planet moves 2,400 times faster than the star, and therefore the planet is 2,400 times less massive than the star. This implied that the planet's mass is about half of the mass of Jupiter, indicative of a planet and not a brown dwarf.

As for Latham's object, the true mass was not measured until 2022, several years after the Nobel Committee had already chosen Mayor and Queloz. The mass of Latham's object was determined with the astrometric method, using data from a space telescope called Gaia to be discussed in chapter 8. As it turned out, the orbital plane is oriented within a few degrees of being exactly perpendicular to our line of sight, just the coincidence that skeptics had warned was possible. The object's true mass is 220 times the mass of Jupiter, making it neither a planet nor a brown dwarf—it's massive enough to

7. Indeed, they used one of the four telescopes officially designated as Very Large Telescopes or VLTs (I'm not kidding). Each has a primary mirror of diameter 8.4 meters. They're located at the European Southern Observatory's facility on Cerro Paranal in northern Chile.

be a full-fledged star. With the benefit of hindsight, the Nobel committee made the right call on that one.

Birkby's 2017 measurement of the true mass of the celebrated 51 Pegasi planet was an important achievement, but it wasn't even close to being the first time an exoplanet's mass was measured unambiguously. That honor belongs to astronomers who were studying a hot Jupiter orbiting a Sun-like star called HD 209458. The special aspect of this system is that our line of sight to the star happens to skim across the plane of the planet's orbit. As a result, the planet's orbit carries it directly in front of the star, just as the Galilean moons cross directly in front of Jupiter. The planet around HD 209458 covers about 1% of the star's surface, causing the star's measured brightness to drop by about 1%.

The dips in brightness of HD 209458 were detected in 1999 by two groups, one led by David Charbonneau of Harvard University, and the other by Gregory Henry of Tennessee State University. The mere existence of these miniature eclipses—known as *transits*—implied that we are viewing the orbit from the side, and thereby eliminated the uncertainty regarding the orbit's orientation (figure 3.6). This circumstance allowed the planet's true mass to be calculated from the Doppler data, as opposed to its minimum possible mass. For this reason, the detection of the transits of HD 209458 was technically a more definitive discovery of an exoplanet than any previous claim, although by 1999, the astronomical community had stopped doubting the Doppler discoveries of the planets around 51 Pegasi and other stars. Still, the transit detection was a pivotal moment in the timeline of exoplanet exploration. It marked the beginning of the second phase of discoveries, which would be based on the transit method.

CHAPTER THREE

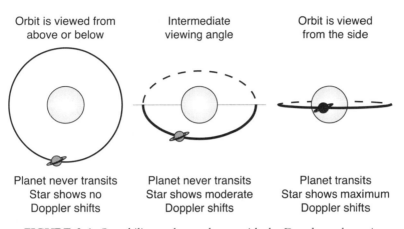

Orbit is viewed from above or below	Intermediate viewing angle	Orbit is viewed from the side
Planet never transits Star shows no Doppler shifts	Planet never transits Star shows moderate Doppler shifts	Planet transits Star shows maximum Doppler shifts

FIGURE 3.6. Our ability to detect planets with the Doppler and transit methods depends on our vantage point. Neither method works when our line of sight is perpendicular to the orbital plane (left panel). For more arbitrary viewing angles (middle panel) there are Doppler shifts, but no transits. Orbits viewed sideways (right panel) show the largest Doppler shifts of any orientation and allow the planet to transit the star.

The Transit Method

On June 8, 2004, there was a memorable eclipse of the Sun. It was not a total solar eclipse, when the Moon blots out the Sun rim to rim, nor even a partial solar eclipse, when the Moon takes a bite out of the Sun and leaves a luminous crescent. In fact, the eclipse wasn't obvious at all. You never would have noticed anything special was happening if an astronomer hadn't told you. I was watching, transfixed, from the roof of one of the buildings at the Harvard College Observatory, using a telescope and a filter that attenuates the Sun's light down to a level that is safe to view by eye. Like the dozens of others attending the roof party, and thousands of other people around the world, I saw a tiny black dot make its way across the Sun

THE AGE OF EXPLORATION

like a pea slowly rolling across a dinner plate. The black dot was Venus making one of its rare passages directly in front of the Sun: a transit of Venus. Although not as spectacular as a solar eclipse, the transit of Venus was fun to watch because such transits are rare[8] and feature prominently in the history of astronomy. In the eighteenth century, astronomers in far-flung locations across the globe saw Venus drift across different parts of the Sun, due to the parallax effect. This allowed them to calculate the distances to Venus and the Sun with improved precision, thereby anchoring their knowledge of the distances and sizes of the objects in the Solar System.

Likewise, our current knowledge about exoplanets is anchored by observations of transits. Even though the stars are so distant that they appear as points of light, and we have no way to watch a planet's silhouette move across the star's luminous disk, we can tell a transit is occurring because the star appears slightly fainter than usual. The planet blocks a small fraction of the starlight that normally reaches us. This happens repeatedly, each time the planet has gone all the way around the star. We can calculate the planet's size by measuring the fraction of the star's light that is blocked by the planet, and we can determine its orbital period by measuring the time that elapses between transits. Then, we can monitor the star's Doppler shift to determine the planet's mass. It bears repeating that we can obtain the planet's true mass and not just the minimum mass, because the occurrence of transits guarantees that we are viewing the orbital plane almost exactly sideways,

8. Because Venus's orbital plane is inclined by 3° relative to Earth's, a special coincidence is required for the Earth, Venus, and Sun to line up closely enough for transits to occur. Transits of Venus occur in pairs separated by 8 years, followed by century-long intervals with no transits. The next ones will take place on December 10–11, 2117, and December 8, 2125.

CHAPTER THREE

like a Frisbee flying horizontally at eye level. In this configuration, the maximum observed Doppler shift reveals the full speed at which the star is being pulled around the center of mass by the planet.

To further appreciate the transit method, imagine that alien astronomers are searching for planets around the Sun from the comfort of their own planetary system, using the same technology available to us. First, they try the Doppler method. By monitoring the Sun's Doppler shift over several decades, they would detect Jupiter, the planet that pulls the Sun the hardest. Jupiter causes the Sun to move around the center of mass of the Solar System with a speed of around 12 meters per second (28 mph), a moderate driving speed. To calculate the fractional wavelength shifts in the Sun's spectral absorption lines, we divide by the speed of light, 299,792,458 m/s, which leads to a result of 0.00000004 (or 4×10^{-8}). Such a minuscule change in wavelength corresponds to a shift of about a hundredth of a single pixel in a typical astronomical camera used to capture a star's spectrum. The shift would nevertheless be detectable, because the data from thousands of different absorption lines can be combined to enhance the signal strength.

The aliens wouldn't be able to detect the Earth, though. The Earth's contribution to the Sun's motion is only 0.09 m/s (0.2 mph), a tortoise's pace. The corresponding shift in the wavelengths of the Sun's spectral lines is 3×10^{-10}, corresponding to 1/20,000th of a single pixel in the digitized spectrum. That's an atomic-scale shift, only a few times larger than the spacing between silicon atoms on the semiconductor surface of the camera's light-sensitive detector. A daunting measurement. Nobody has yet been able to achieve such a high precision. For now, detecting truly Earth-like planets with the Doppler method is beyond our capabilities.

THE AGE OF EXPLORATION

Now, imagine that the aliens are fortuitously located somewhere in the galaxy where they have a side view of the Solar System. When Jupiter is in front of the Sun, the Sun appears to fade by 1% for about one day, before going back to its usual brightness. Although that might seem like a small effect, measuring a 1% change in brightness is a cinch in comparison to measuring a fractional Doppler shift of 4×10^{-8}. The aliens could detect a 1% change using a telescope small enough to hold in their hands (or tentacles). The transiting Earth, on the other hand, would cause the Sun to fade by only 0.01%. Such a tiny change cannot be detected with a backyard telescope, but it *can* be detected with one of today's space telescopes. In this sense, the transit method is currently superior to the Doppler method. It's the only method that gives us a fighting chance to detect Earth-sized planets at Earth-like orbital distances around Sun-like stars with today's technology.

The transit method has another advantage over the Doppler method. It's possible to monitor the brightness of hundreds of thousands of stars at the same time by using a wide-angle camera. In contrast, measuring precise Doppler shifts is almost always done one star at a time. We can only handle one star at a time because we need to measure the intensity of the starlight at each of approximately 100,000 different wavelengths, which usually requires spreading out the star's spectrum across the entire detector, leaving no room for the spectrum of another star.

These virtues of the transit method are accompanied by a tragic flaw. From a single vantage point in the galaxy, only a small percentage of exoplanets are ever seen to transit their stars. The geometric probability that the line of sight of a randomly located observer skims closely enough to a planet's orbital plane to see transits turns out to be equal to the radius of the star divided by the radius of the orbit.[9] If there were alien astronomers viewing the Solar System from every possible direction, only about one out of 200 would be appropriately positioned to see the Earth transiting the Sun. Likewise, whenever we detect a planet orbiting at 1 AU around a nearby Sun-like star, we can be pretty sure there are hundreds of similar planets around other stars that are undetectable with the transit method.

A related problem is that even when transits do occur, they're brief. If the aliens observed Jupiter transiting the Sun, the transit would last about a day, which might make detecting Jupiter sound like quick work—until you remember that Jupiter's orbital period is 12 years. To have a reasonable chance of detecting Jupiter, the aliens would need to monitor the Sun continuously for years, waiting for the one day when the Sun fades by 1%. They'd better hope it's not cloudy that day.[10] Even after they detected the dip in the Sun's brightness, they'd need to keep monitoring for another dozen years to measure Jupiter's orbital period. They'd probably want to wait yet another dozen years to see if a third transit happens at the expected time and confirm that the first two dips weren't merely glitches in their detector.

---

9. To be more precise, the geometric probability for transits to occur is $\dfrac{R}{a(1 - e^2)}$, where R is the star's radius, a is the semimajor axis, and e is the eccentricity.

10. That's another reason why astronomers love space telescopes. From space, you can observe for months without being interrupted by bad weather, or by the Sun rising and spoiling the view of the stars.

THE AGE OF EXPLORATION

Serendipity and Prophecy

Throughout this chapter, I've been moaning and groaning about how hard it is to detect exoplanets. I'd like to end on a different note. Despite the technological challenges, and despite all the false starts, the discovery of exoplanets was a rare and wonderful occasion when a scientific endeavor turned out to be *easier* than expected. Usually, Murphy's Law prevails: things go wrong and tasks prove to be more difficult than expected. In this case, the job was easier because of the existence of hot Jupiters, an unexpected gift from Mother Nature. Hot Jupiters produce much larger Doppler signals than would be produced by any direct analogs of the planets in the Solar System. Because of their tiny orbits, hot Jupiters are also more likely to transit. The geometric transit probability of a hot Jupiter is on the order of 10%, as compared to 0.1% for a planet at Jupiter's orbital distance. Because of their short orbital periods, hot Jupiters can be detected and confirmed with only a few weeks' worth of data. There's no need to patiently accumulate data over decades, as Gordon Walker thought would be necessary when he set out to find planets in the 1980s. The only drawback of hot Jupiters is that they're rare. Only about 0.5% of Sun-like stars have a hot Jupiter. A major reason why Mayor and Queloz succeeded is that they were granted enough telescope time to search hundreds of stars and overcome the low odds.

Although the discovery of hot Jupiters came as a surprise, it's not quite true that *nobody* foresaw them. In 1952, Otto Struve, an astronomer at the University of California at Berkeley, published a short paper pointing out that the precision of Doppler measurements had become good enough to detect planets—but only if there existed planets at least as massive as

Jupiter with orbital periods as short as a few days. Setting aside the question of how such a planet might have formed, he realized there is no law of physics that forbids such planets from existing. In an alternate history, Struve's paper inspired astronomers to launch a thousand ships and explore nearby stars for hot Jupiters. In fact, his paper languished in obscurity. None of the pioneers—neither Walker, Latham, Mayor, nor Queloz—were influenced by Struve's paper. The planet around 51 Pegasi probably could have been discovered in the early 1960s, or surely by Walker in the 1980s, had the Telescope Time Allocation Committee allowed him to observe a larger number of stars.

Planet formation theory, and our expectations based on the Solar System, turned out to be like medieval maps of the world: accurate in the vicinity of home, while also exaggerating the size of some landforms, minimizing others, and missing entire continents. I try to remember the history of the discovery of exoplanets, with all its twists and turns, when I am suffering from excessive pessimism. Even when a research topic has been tainted by claims that turned out to be wrong, even when talented people have already looked and found nothing, even when theorists think your idea is farfetched, there could still be a spectacular discovery waiting to be made.

THE AGE OF EXPLORATION

CHAPTER FOUR

HERE, THERE BE GIANTS

The theory of planet formation is a rags-to-riches story, with lowly grains of dust in the opening act, and fully formed planets in the finale. Written and rewritten by many authors over several centuries, the theory is a complex drama with a large cast of physical principles, ranging from the aerodynamics of dust grains to the thermodynamics of planetary collisions. In the light of exoplanet discoveries, is this drama based on a true story? Or is it a fantasy?

The answer is not fully known, but surely lies somewhere between fact and fiction. So far, I have emphasized the contradiction between the theory's predictions and the earliest exoplanet discoveries. In this chapter, which concentrates on giant planets, we will encounter more contradictions, but we will also see that some predictions were borne out. In the next chapter, which concentrates on smaller planets, we will meet exoplanetary systems that had no role in the theoretical drama. They were neither predicted nor were they forbidden by the theory.

One aspect of the theory has held up reasonably well. Both in theory and in reality, nature seems to distinguish between giant planets and smaller planets, with a dividing line at around 4 $R_\oplus$, four times the radius of the Earth. One of the sharpest differences is that the inner regions of planetary systems are more likely to be populated by small planets than giant planets. Suppose you chose a hundred Sun-like stars at random and tallied up all the planets around those stars with orbits smaller than 1 AU. On average, you'd find about 95 planets with sizes between 1 and 4 $R_\oplus$, with some stars having no such planets, and other stars having as many as six. In contrast, you'd find only about a dozen planets larger than 4 $R_\oplus$, with few if any stars having more than one. So, while the mere existence of hot Jupiters and other close-orbiting giant planets was a shocking discovery, we should keep in mind that they are uncommon.

Another intriguing difference between giant planets and small planets is that the probability for a star to have a giant planet depends on the chemical composition of the star, while for smaller planets, any such dependence is weaker. Stars are composed almost entirely of hydrogen and helium, but they also contain a smattering of heavier elements, and the heavy-element fraction varies from star to star. An average kilogram of material drawn from the Sun's atmosphere consists of 740 grams of hydrogen, 240 grams of helium, and 20 grams of heavier elements such as carbon, nitrogen, oxygen, and iron. Scooping a kilogram from 51 Pegasi instead of the Sun, we would find about 30 grams of heavy elements, while a kilogram of Arcturus's atmosphere would contain only 6 grams of heavy elements. The reason for the variation from one star to another is that heavy elements were not made in the Big

Bang. They were forged by exotic and often catastrophic astronomical events, such as supernova explosions. These events occur randomly, causing heavy elements to be spread unevenly throughout the galaxy. Heavy elements also accumulate over time as more of these exotic and catastrophic events take place. As a result, stars have different abundances of heavy elements depending on when and where they formed.

We can determine a star's composition based on the pattern of absorption lines in its spectrum—the same lines that are used to track a star's Doppler shift. As it turns out, whenever a star's spectrum of light indicates that the star is relatively rich in iron, carbon, oxygen, and other elements heavier than hydrogen and helium, we're more likely to detect a giant planet around that star.

This effect is so large and unmistakable that it was noticed early in the history of exoplanetary science, in 1997, by Guillermo Gonzalez, an astronomer then at the University of Texas at Austin. A solar-mass star consisting of 2% heavy elements by mass, like the Sun, has a one-in-twenty chance of having a Jupiter-mass planet with an orbit smaller than a few AU. For an otherwise identical star with 6% heavy elements, the odds rise to about one in four. The rise in giant-planet production associated with an increase in a star's heavy-element inventory is called the "metallicity effect." The name was chosen because— by a convention that is hard to explain or defend—astronomers refer to any element besides hydrogen and helium as a metal. Even decidedly non-metallic elements such as carbon, oxygen, and chlorine are, in astronomical lingo, metals.[1]

---

1. This recalls the stories (legends?) of primitive societies in which the only words for numerical quantities are *one*, *two*, and *many*. For astronomers, the periodic table consists of *hydrogen*, *helium*, and *metals*.

The metallicity effect is circumstantial evidence supporting the core accretion theory of giant planet formation. Nobody predicted the metallicity effect prior to its discovery, but in retrospect, it's a logical outcome of the events described in the theory. If the gaseous cloud that collapses to form a star is unusually rich in heavy elements, then there will be more dust and ice strewn about the star's protoplanetary disk. A higher abundance of dust and ice, in turn, makes it easier for a growing planet to reach the threshold mass of 8 $M_\oplus$ that initiates the runaway accretion of hydrogen and helium and the formation of a giant planet. Based on the same logic, we would not expect a strong metallicity effect for small planets, because there is no threshold mass to form a small planet; they can form even in relatively metal-poor conditions. And, indeed, the probability for a star to have planets smaller than 4 $R_\oplus$ (about the size of Uranus or Neptune) is observed to be nearly independent of the star's chemical composition.

Furthermore, both in theory and in reality, Earth-mass exoplanets are mainly solid bodies, and Jupiter-mass exoplanets are mainly gaseous. In theory, this is because all planets start out as small and solid, and the only way to grow into a giant planet is to swallow enormous quantities of hydrogen and helium gas from the protoplanetary disk. In reality, we can tell whether an exoplanet is mainly solid or gaseous by calculating the planet's overall density: we measure its mass (with the Doppler method) and radius (with the transit method), and then divide the mass by the volume. An Earth-mass ball of gas would have a lower density than an Earth-mass ball of rock and metal. And, indeed, we have never found an Earth-mass ball of gas, nor have we found a Jupiter-mass planet with a density high enough to be compatible with a ball of rock and metal. I'll have more to say about the masses and densities of exoplanets in chapter 5. For now, let's just score another couple of points for the theory.

Now for the theory's shortcomings. Planetary orbits were expected to be nearly circular and to lie within a single plane, and gas giant planets were supposed to be excluded from the inner few AU surrounding the star. We have found exoplanetary systems that violate each of these basic theoretical expectations.

The most flagrant disagreements between theory and observations pertain to giant planets, which are between 4 and 20 $R_{\oplus}$ in size (for reference, Jupiter's radius is 11 $R_{\oplus}$). The mere existence of giant planets with orbits smaller than a few AU was forbidden by the theory because giant planets are only supposed to be able to form beyond the snow line, where the conditions are favorable for core accretion. Nevertheless, in reality, approximately 5% of Sun-like stars do have a Jupiter-mass planet within 1 AU. This was Surprise No. 1 of exoplanetary science.

Hot Jupiters are the most extreme cases of these "misplaced giants." The first known hot Jupiter was 51 Pegasi b, the planet whose discovery initiated the exoplanet revolution in 1995. The "b" in 51 Pegasi b indicates that we're talking about the planet, and not the star. The usual naming convention for planets is to begin with the star's name and add "b" for the first discovered planet, "c" for second planet, and so forth. The letter "A" is reserved for the star itself, although it is usually dropped unless there is any danger of confusion. Hot Jupiters, roughly defined as planets with orbits smaller than 0.1 AU and masses of at least one-tenth the mass of Jupiter, occur around about 0.5% of Sun-like stars. So, they are rare—but they still demand an explanation. Often, the most extreme cases are the most revealing.

CHAPTER FOUR

An Egg-Shaped Exoplanet

One of the most extreme cases is a hot Jupiter called WASP-12b, the twelfth exoplanet discovered by the Wide-Angle Survey for Planets, which is located 1,400 light-years away in a direction between the constellations of Gemini and Auriga. Plate 6 is an artist's conception of the planet and its nearby star. We see an egg-shaped planet, with orange and brown horizontal bands on one side. The other side of the planet, facing the star, is too bright to make out any details. The planet is connected to the star by a stream of luminous material that spirals around the star to form a Saturn-like ring.

Oh, how I wish this were a real image . . . but we have absolutely no way to resolve such fine spatial details of an exoplanetary system. Instead, our most direct knowledge of WASP-12b can be visualized in a pair of plots (figure 4.1). The upper plot displays the star's brightness versus time, spanning three orbits of the hot Jupiter. The star fades by 1.3% during each of the three transits, implying that the planet's cross-sectional area is 1.3% of the cross-sectional area of the star. Given this measurement, and our best estimate of the star's radius, we can calculate the planet's radius, which comes out to be about 20 $R_\oplus$. That's nearly twice Jupiter's radius. WASP-12b is not merely a giant planet—it's a swollen planet, larger than theoretical models of planets are able to explain. A ball of mainly hydrogen and helium with the mass of WASP-12b should contract under its own gravity until it is approximately the size of Jupiter.

What caused WASP-12b to balloon in size? We don't know. WASP-12b is a member of a subcategory of planets called "inflated hot Jupiters," of which dozens are known. They tend to be the hottest of the hot. The main risk factor for a

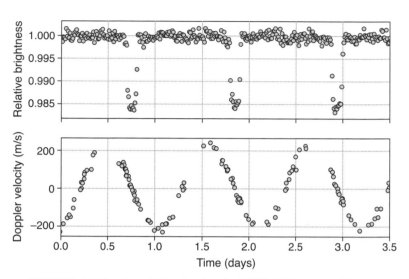

FIGURE 4.1. Transit and Doppler data for WASP-12. Top panel: The star fades in brightness by 1.5% every 1.1 days, when the planet moves in front of the star. Bottom panel: The planet's gravity causes the star to circle around the center of mass at a speed of 220 meters per second.

hot Jupiter to become inflated seems to be an atmospheric temperature exceeding 1500°C, which happens when the planet has an unusually short orbital distance or orbits an unusually hot star. Given this trend, it seems likely that inflation is caused by extreme heat. After all, heat causes gases to expand. However, to swell up an entire planet to twice its ordinary size, the heat would need to be transferred from the planet's atmosphere to its interior, and it has proven difficult for theorists to explain how the heat from the star could penetrate deeply enough within the planet. This quandary is called the "radius inflation problem." In standard models of giant planets, the energy from starlight is either reflected, or absorbed in surface layers and then quickly lost into space via the planet's own infrared radiation.

CHAPTER FOUR

Setting aside the radius inflation problem, let's return to figure 4.1. The lower plot shows Doppler measurements of the speed of the star WASP-12 toward or away from the Solar System. The data points rise and fall in a pattern that repeats every 1.1 days, telling us that the planet's orbital period is 1.1 days. The measured speed varies by as much as 220 meters per second away from the average speed, from which we can calculate that the planet is 40% more massive than Jupiter. Furthermore, the data points rise and fall in a gentle wavy pattern called a sinusoidal function, from which we can deduce that the orbit is nearly circular. The Doppler measurements for WASP-12 are good enough to rule out an orbital eccentricity larger than 0.05, making the planet's orbit at least as circular as Jupiter's and Saturn's orbits around the Sun.

In our day-to-day research, we pore over plots like those shown in figure 4.1. Engrossed by a challenging measurement or a technical calculation, I often conceive of exoplanets as patterns of data points on a plot, or as solutions to celestial physics problems, not as picturesque alien landscapes. So, it's refreshing and stimulating when artists help us visualize an exoplanetary scene. In the case of WASP-12, the artist was well informed. Some of the details in the picture are based on reasonable inferences we have drawn from the data and our knowledge of physics. Only a few details are mistaken or wholly made up.

For example, the image correctly portrays the planet as about one-tenth the size of the star. However, the planet in the picture is too close to the star. The actual orbital radius is three times larger than the star's radius. Although that's a larger orbit than in the picture, it's still a surprisingly small orbit by astronomical standards. It would have been a lot easier for your elementary school class to make a scale model of the

HERE, THERE BE GIANTS

WASP-12 system than the Solar System. Instead of being spread out over the entire playground, the WASP-12 model would have fit on the teacher's desk, with a grapefruit for the star and a grape tomato for the planet.

Another faithful detail in the picture is the intense glow coming from the planet's *dayside*, the side of the planet facing the star. The dayside of WASP-12b is getting broiled. Based on the star's power output, and the distance from the planet to the star, we can calculate that the planet's dayside is heated to about 2300°C. The planet's *nightside*, facing the blackness of space, is at least several hundred degrees cooler than the dayside.

The narrow orange and brown bands within the planet's atmosphere are a bit of artistic license. The artist, understandably, chose to make the planet resemble Jupiter. What would it *really* look like? Hot Jupiters have attracted the attention of experts in computer simulations of global circulation patterns—the same kinds of simulations that are used to model the Earth's climate as well as explain the bands and whorls in the atmospheres of Jupiter and Saturn. According to the simulations, the intense heating of the dayside of a hot Jupiter should drive a powerful wind that circulates around the planet's equatorial zone. The wind is predicted to blow in the same direction as the planet is rotating, with a speed of several kilometers per second (thousands of miles per hour). At higher latitudes, the wind is expected to blow toward the poles on the dayside and curl back toward the equator on the nightside, possibly creating giant vortices. There would be, at most, only two or three latitudinal bands with different wind patterns, as compared to the dozen or so bands seen in Jupiter's atmosphere. Soon, we might be able to test the basic predictions of hot Jupiter atmospheric models against observations, as discussed in chapter 6.

CHAPTER FOUR

What about the planet's oval shape? You might be surprised to learn that this detail in the picture is true to life. The planet's shape cannot be observed directly, but we know it is being squashed by *gravitational tidal forces*.

The word "tidal" brings to mind the ebb and flow of the ocean tides, but in astronomy, tides are a more general phenomenon. According to Newton's law of gravity, the gravitational force exerted by a star on a planet weakens as the inverse square of the distance between them. The weakening of the gravity with distance causes the planet's dayside to feel a stronger gravitational force than the nightside, because the dayside is closer to the star than the nightside. The *differences* in the gravitational attraction experienced by different parts of the planet are called tidal forces. Tidal forces are capable of some impressive feats. They can squash a planet, alter its rotation, mold its orbit into a different shape, and in extreme cases, tear it to shreds.

Imagine that we grab a planet, step 1 AU away from a star, and let the planet go, starting from rest. What happens next? The planet falls toward the star due to the star's gravitational force. At any point along the way, the planet's dayside is pulled harder than its nightside. This causes the dayside to accelerate faster than the nightside. The initially spherical planet is stretched out into an ellipsoid (a three-dimensional ellipse), with the long axis pointing toward the star (figure 4.2). Less obviously, the same stretching effect occurs if we throw the planet sideways, putting it into orbit around the star instead of letting the planet fall directly toward the star. The sideways motion doesn't change the fact that the planet's dayside is being pulled harder than the nightside. The orbiting planet gets stretched into an ellipsoid whose major axis always points toward the star. According to calculations by Victoria Antonetti

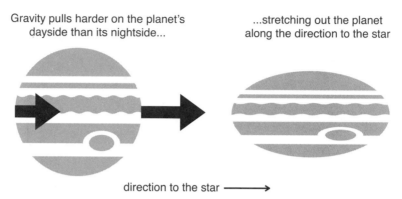

Gravity pulls harder on the planet's dayside than its nightside...

...stretching out the planet along the direction to the star

direction to the star ⟶

FIGURE 4.2. Gravitational tidal forces tend to stretch a planet along the direction to the star. To a lesser degree, the star is also affected by tidal forces from the planet.

and Jeremy Goodman of Princeton University, the diameter of WASP-12b is about 10% longer when measured in the direction pointing toward the star than in a perpendicular direction. In the most dramatic known case, a terrestrial-type planet called KOI 1843.03 is being stretched into the shape of an American football, according to calculations by Ellen Price and Leslie Rogers at the University of Chicago. Depending on the planet's composition, the length of its longest diameter (from tip to tip of the football) is somewhere between 30% and 80% longer than the length of its shortest diameter (across the football's circular cross section).

Because tidal forces depend only on *differences* in force across the extent of the planet, rather than the overall force on the planet, the strength of tidal stretching varies inversely as the *cube* of the distance to the star, rather than the square.[2] As a

2. For math aficionados: this is because $\frac{1}{(a-d)^2} - \frac{1}{a^2} \approx \frac{2d}{a^3}$ when d is much smaller than a. Here, a is the distance from the planet to the star, and d is the planet's diameter.

result, tidal forces are substantial only when two astronomical bodies are very close together. The planets in the Solar System are too far from the Sun for tides to alter their shapes by very much. Tides from both the Sun and the Moon cause the Earth's oceans to slosh up and down by a few meters, on average, which is important for sailors, marine biologists, and beachgoers, but is negligible on a planetary scale.

The story is different for hot Jupiters and other close-orbiting exoplanets, for which tidal stretching can be severe. The stretching is resisted by the planet's own gravity, which attracts all the material composing the planet toward the planet's center. A solid planet also resists being stretched by the internal strength of rock and metal.

If we could move a planet closer to its star, the tidal forces would increase in magnitude and eventually overwhelm the forces holding the planet together. The planet would be destroyed. A terrestrial planet would crack into smaller pieces, and a gaseous planet would be shorn apart. Streams of gas would erupt from a gaseous planet and expand into the surrounding space or spiral down onto the star, as in the artist's conception of WASP-12b. The minimum safe distance from a star that allows a planet to maintain its integrity, despite the stretching effect of tidal forces, is called the *Roche limit*, named after Édouard Roche, a French theoretician who published a treatise on this subject in 1849. The numerical value of the Roche limit depends on the planet's composition. WASP-12b, with an orbital radius of 0.023 AU, is very close to the limit. The planet is peering over the precipice.

Even more interesting is that WASP-12b is moving closer and closer to the edge. By precisely timing the transits of the planet, my research group and others have shown that the orbital period is decreasing, and therefore, according to Kepler's third law, the size of the orbit is shrinking. In 2009, when the planet was discovered, the orbital period was 1.091421 days, but by 2020, the period had fallen to 1.091418 days, an average decrease of 29 milliseconds per year. At that rate, the planet will violate the Roche limit and suffer the fatal consequences within a million years—basically, tomorrow, by astronomical standards.

What is causing WASP-12b to spiral inward? It's probably tidal forces, again, but this time the relevant effect is not the star squishing the planet, but rather, the planet squishing the star. Newton's third law strikes again ("for every action . . ."). Because of the star's much larger mass, it is not distorted nearly so much as the planet, but the tidal forces still have important long-term consequences. As the planet revolves around the star, the tidal force on the star constantly changes direction, setting the material within the star into motion. The details of the resulting oscillations within the star are complicated, but the expected result is that the star, which was initially spinning slowly, should start spinning faster in the direction of the planet's orbital motion. It's as though the star's surface were being tugged around by the orbiting planet. If the star spins faster, its angular momentum increases, and the planet's angular momentum must decrease to compensate. The planet gradually transfers its angular momentum to the star, causing its orbit to shrink.

The orbit of the Moon is changing, too, due to tidal interaction with the Earth, but because the Moon revolves slowly (once per month) compared to the Earth's rotation (once per day), the effects are to *slow down* the Earth's rotation and *widen* the orbit, the opposite of the effects in the WASP-12 system. That's why the Moon is observed to be receding from the Earth at a rate of 3.8 centimeters per year. This astonishing fact was established in a series of beautiful experiments in which astronomers fired laser pulses at the Moon. A small portion of the laser light was reflected from mirrors installed by the Apollo astronauts. By monitoring the round-trip times of the pulses, the astronomers could measure tiny changes in the Earth-Moon distance.

Eccentric Exoplanets

WASP-12b epitomizes Surprise No. 1 of exoplanetary science, the existence of gas giant planets on tiny orbits. To introduce Surprise No. 2, I direct your attention to a Sun-like star named HD 80606, located 217 light-years away in the constellation of Ursa Major (the constellation that houses the Big Dipper). In 2001, a team of astronomers found the star to be wobbling back and forth, using the Doppler technique. However, the pattern of Doppler shifts (figure 4.3) was unlike the gentle wavy pattern that was seen for 51 Pegasi and WASP-12. For an interval of about 100 days, the star moves toward us, gradually increasing its speed by about 3 meters per second with each passing day. Then, suddenly, the star jams on the brakes and throws itself into reverse, jolting away from us with an acceleration of about 1,000 meters per second per day. After a couple of days, the star resumes accelerating toward us, and the pattern repeats.

The explanation for this peculiar behavior is that the orbit of HD 80606b is not even remotely circular. It's a highly elongated ellipse. The planet, obeying Kepler's second law, moves sedately when it is far from the star and maniacally when it is close to the star. The star, obeying Newton's third law, responds by moving around the system's center of mass in the opposite

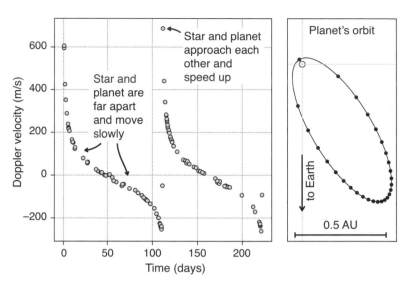

FIGURE 4.3. The HD 80606 system. The left panel shows that the star's speed changes dramatically throughout an orbit. The right panel illustrates the planet's highly elliptical orbit. Black dots show the planet's location at 4-day intervals. The rapid rise in the star's speed occurs when the planet is closest to the star.

direction as the planet. When the planet whips around behind the star, the star speeds up in response, producing the spike we see in the Doppler data.

Ponder for a moment this extraordinary plot of Doppler velocity versus time, and its implications for the planet. Here on Earth, the seasonal variations in climate are caused by the 23.5° tilt of the Earth's rotation axis with respect to its orbital axis, and not by the Earth's modest orbital eccentricity of 0.017. On HD 80606b, the effects of any axial tilt would be overwhelmed by the much stronger effects of the changing distance between the planet and the star. Imagine looking at the sky from the planet's surface (disregarding, for imagination's sake, the complication that it's a gas giant with no solid surface to stand on). For most of the planet's 112-day orbit, the star is about the same size in the sky as the Sun appears in our sky. During the week-long summertime, when the planet approaches the star at high speed, the star looms larger and larger, swelling to almost 30 times its usual size. On Earth, you can block your view of the Sun's disk with a pea held at arm's length. In midsummer on HD 80606b, you'd need to use a dinner plate instead of a pea. And it'd be hard to find a plate that can withstand the heat. The planet's temperature rises to a maximum of about 1100°C before the onset of spring and the retreat of the planet to a more Earth-like distance.

To physics teachers, HD 80606b is useful as a vivid demonstration of Kepler's first and second laws of planetary motion. To planet formation theorists, HD 80606b is a provocation. Its orbital eccentricity of 0.93 is much higher than would be expected of a planet that formed within a circular protoplanetary disk. On a plot in which the horizontal axis is the orbital period, and the vertical axis is the orbital eccentricity (figure 4.4), HD 80606b stands tall, with one of the most extreme eccentricities of all known planets. The plot also illustrates a more general point: the known exoplanets have a wider range of orbital eccentricities than the planets in the Solar System. This was Surprise No. 2 of exoplanetary science.

The plot shows another interesting pattern: the region near the upper left corner is nearly empty. Evidently, the planets with orbital periods shorter than 10 days tend to have lower eccentricities than longer-period planets. This pattern, at least, we understand. Even though we don't yet know what causes planets to *attain* high orbital eccentricities, we do know how short-period planets *lose* their eccentricities. It's tides, again.

HERE, THERE BE GIANTS

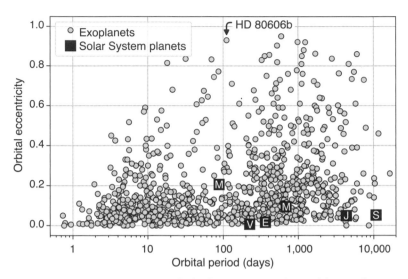

FIGURE 4.4. Measurements of orbital eccentricity and period for exoplanets detected with the Doppler method, and for Mercury, Venus, Earth, Mars, Jupiter, and Saturn. The exoplanets have a wider range of eccentricities.

The amount by which tidal forces distort a planet's shape depends sharply on the distance from the star. A planet on a highly eccentric orbit gets squished more when it is close to the star, and less when it is far from the star. Each time the planet approaches the star closely and then recedes, the star's tidal grip strengthens and weakens like a fist clenching and unclenching around a ball of clay. This causes the planet to heat up. The internal friction within the planet converts some of the squeezing motion into heat, just as it does for the ball of clay. The result is that the energy associated with the planet's orbital motion is slowly degraded into heat. Because of the loss of orbital energy, when the planet recedes to the far side of its orbit, it falls slightly short of the maximum distance that it achieved in the previous orbit. The

CHAPTER FOUR

length of the semimajor axis of the ellipse shrinks; the orbit becomes smaller and slightly more circular. Given enough time, the orbit becomes a small and nearly perfect circle. Once that happens, the distance to the star no longer changes with time, and there is no more periodic squeezing. The loss of orbital energy ceases. Thus, given the opportunity, tidal forces can take a wide eccentric orbit and mold it into a tiny circular orbit.

Hold that thought for another few pages. It's the kernel of a possible solution to the mystery of hot Jupiters.

Tilted Orbits

We've discovered giant planets in locations that should have been too hot for giant planets to form. We've found orbits that are too eccentric to have emerged from a circular protoplanetary disk. What about the third main expectation of planet formation theory—the good alignment between the planes of different planetary orbits? Are exoplanetary systems flat, like the Solar System, or should we brace ourselves for another surprise?

Obtaining this information for giant planets is difficult because almost all the known systems with more than one giant planet were detected with the Doppler method. The Doppler method allows us to detect multiple planets around the same star; in such cases, the star's Doppler shift is the sum of the shifts caused by each planet. However, the Doppler method doesn't reveal the angles between the planets' orbits, because of the usual limitation that we can only track the star's radial velocity as opposed to its full three-dimensional velocity. With only Doppler data, we can't tell if two planets are revolving in the same direction or opposite directions, or if their orbits are tilted by an angle between these two extremes.

It turns out to be easier to measure the angle between a planet's orbital plane and the star's equatorial plane—or, equivalently, the angle between the planet's orbital axis and the star's rotation axis. The Sun's rotation axis and the Earth's orbital axis are nearly parallel, differing in direction by only 7°. In astronomical jargon, the Sun's *obliquity* is 7°, a small angle, consistent with the close alignment of the planets' orbits. How could we possibly measure the obliquity of a distant star? It sounds impossible because, as I've stressed, stars almost always appear as structureless points of light, even with our best telescopes. How can we sense a star's rotation if we cannot make out any details on the star's surface?

The solution to this problem is a clever trick based on a fusion of the Doppler and transit methods. During a transit, a planet passes in front of its star, and although we can't see the planet's shadow moving across the star's surface, we know it's happening because we can detect the decrease in the star's brightness. Now imagine watching the same event with high-tech glasses that allow you to perceive tiny Doppler shifts (plate 7). Through these glasses, the star looks weird: one half of the star is red, and the other half is blue. This is because the star is rotating. The side of the star that is spinning away from the Earth is redshifted, and the side that is spinning toward us is blueshifted. When we observe this star with a telescope, as opposed to high-tech glasses, the star appears as a single point of light; the light from both sides of the star blends together in our spectrograph. The redshift and the blueshift cancel each other out, leaving no net Doppler shift.

During a planetary transit, though, the planet breaks the symmetry. At first, the planet moves in front of the approaching half of the star, blocking some of the blueshifted light and causing the total starlight to appear slightly redshifted. In the latter half of the transit, the planet covers a small portion of the star's receding half, blocking some of the redshifted light and causing our spectrograph to register a net blueshift.

The sequence of events I just described would be observed for a star with a *low* obliquity. That's the case in which the planet begins a transit in front of the star's approaching side and moves toward the receding side. If, instead, the star's equator is tilted away from the planet's orbital plane, we would see a different pattern. Depending on the angle, the planet might spend less time covering the blueshifted side than it does covering the redshifted side, and the duration of the anomalous redshift would be shorter than that of the blueshift. Or vice versa, with a longer redshift than a blueshift. If the obliquity were nearly 90°, then the planet might spend the entire transit in front of the redshifted side of the star, and we would observe an anomalous blueshift throughout the entire event, with no corresponding redshift. The bottom line is that by tracking the star's apparent Doppler shift throughout a transit, we can measure the angle between the planet's trajectory and the star's direction of rotation.

I've been pursuing these types of measurements since 2005, but I can't take credit for the idea. Like many techniques in exoplanetary science, it was dreamed up long ago by astronomers who were studying the eclipses of stars by other stars, rather than by planets. Many stars have companion stars, and sometimes these pairs of stars (*binary stars*) eclipse each other as they orbit around their center of mass. The first conclusive observations of the anomalous Doppler shifts of eclipsing binary stars were reported in 1924 by Richard Rossiter and Dean McLaughlin, astronomers at the University of Michigan. This phenomenon is now known as the Rossiter-McLaughlin effect.

By exploiting the Rossiter-McLaughlin effect, my research group and others have measured the obliquities of stars in hundreds of exoplanetary systems. The results are wonderful. Just as Surprise No. 1 of exoplanet science was the existence of close-orbiting giant planets, and Surprise No. 2 was the broad range of orbital eccentricities, Surprise No. 3 was the broad range of stellar obliquities. Although many stars have low obliquities—some of them even lower than the Sun—there are glaring exceptions. Our highly eccentric friend, HD 80606b, has an orbit that is tilted by about 40° relative to the star's equatorial plane. Other planets have orbits that carry them nearly above the north and south poles of the star, such as WASP-7b, with a measured angle of 85°. There are also some bizarre systems that appear to be backward, with the star rotating one way and the planet revolving in the other direction. One of the best studied systems is called K2–290. It has two planets whose orbital planes are aligned to within 10°, and the star is turned upside down, with an angle of 124° between its spin axis and the planets' orbital axes.

Tidally Driven Migration

In textbooks, science is said to work in the following way. Scientists make a hypothesis. Then, they make observations to gather evidence for or against the hypothesis. Failed hypotheses are rejected, and successful hypotheses are synthesized into a theory, or, to put it more grandly, a paradigm. Paradigms are always tentative. When new things are discovered that don't fit the theory—such as misplaced giant planets on misshapen and misaligned orbits—the theory is ruled out. The paradigm is shattered.

The adherents of the core-accretion theory of giant planet formation have not followed the textbook procedure. The theory was allowed to marinate in the brains of astronomers for a long time without any new observations, besides the long-established facts of the Solar System. Even after being bashed by new and surprising observations, the paradigm did not shatter. It has managed to survive even the most grotesque violations of our expectations about exoplanets.

The way theorists responded is not by ripping up the script of planet-formation theory and starting over. They preserved the setting, costumes, main characters, and Acts I and II, and added new plot twists at the very end of Act III. Theorists realized that nobody had been paying enough attention to what might happen to planets *after* they form. They have spent the years since 1995 investigating many processes that might alter planetary orbits, in the hope of explaining the exoplanet data. Within the year following the discovery of 51 Pegasi b, two different hypotheses were proposed to explain how a giant planet could attain an orbit with a radius of 0.05 AU, even if it originally formed beyond the snow line at about 3 AU.

When trying to shrink a planet's orbit, the most basic question is how to get rid of most of the planet's orbital energy. Only by losing orbital energy can an orbit's semimajor axis be decreased, bringing the planet closer to the star. The semimajor axis can be reduced still further if the planet somehow loses a lot of angular momentum. The two hypotheses proposed in 1996 differ in how the planet sheds its energy and angular momentum.

The first scenario requires several giant planets or other massive bodies to play starring roles, absorbing some of the energy and angular momentum. Imagine, for example, that the protoplanetary disk has enough material to form three or more giant planets. These giants start out beyond the snow line, on nearly circular orbits that are aligned with each other and with the star's equator. For a while, the planets revolve around the star in what seems like a harmonious arrangement. After a while, it becomes apparent that their orbits are a little too close for comfort. Over billions of years, the configuration proves to be unstable.

In addition to the star's gravitational force on each planet, the planets exert gravitational forces on each other. These interplanetary forces allow the planets to trade energy and angular momentum, altering their orbits. The orbits' eccentricities and inclinations start drifting away from their initial values. Occasionally, gravitational interactions between planets lead to close encounters and near-collisions, during which the gravitational forces can slingshot the planets around in random directions. One planet might get slowed down during the encounter, causing it to fall from a circular orbit into an elliptical orbit that dives close to the star. Another planet might emerge from a near-collision with a speed high enough to escape the planetary system altogether; such a planet would sail away into the interstellar void.

All the neat and tidy properties of the initial system can be lost due to interactions and encounters between giant planets, if the planets are massive enough and if their orbits were initially close together. These effects might help to explain the eccentric and misaligned orbits.

By adding a final plot twist, we might also be able to explain the existence of hot Jupiters. Suppose that by chance, one of the giant planets ends up on a highly eccentric orbit, with one end near its initial location far from the star, and the other

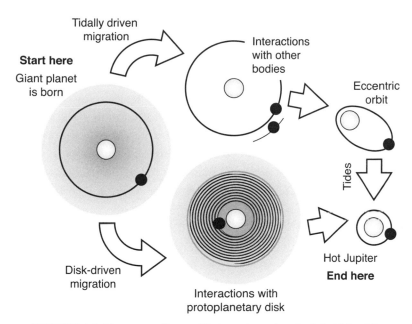

Tidally driven migration

Start here

Giant planet is born

Interactions with other bodies

Eccentric orbit

Disk-driven migration

Interactions with protoplanetary disk

Tides

Hot Jupiter

End here

FIGURE 4.5. Two ways a "normal" giant planet (on the left side) might shrink the size of its orbit and become a hot Jupiter (lower right). Adapted from an illustration by R. Dawson, Penn State University.

end located very close to the star. As is the case for HD 80606b, the planet would periodically make close approaches to the star. During these close encounters, the planet would feel strong tidal forces, and as noted earlier, the combination of tidal forces and friction would cause orbital energy to be dissipated as heat, eventually shrinking and circularizing the orbit. This is one possible way to convert an ordinary Jupiter into a hot Jupiter (figure 4.5).

Theorists have come up with many variations on this scenario. For example, it's not necessary to invoke sudden and close encounters between planets. The effects of gravitational interactions between planets can accumulate gradually, over millions of years. In another variation, the star at the center of the planetary system has a companion star located hundreds or thousands of AU away, whose gravitational pull is feeble but nevertheless causes gradual changes in the planets' orbits. As noted earlier, many stars do have companion stars. More than half of the points of light in the night sky are multiple-star systems for which the stars are so closely spaced that our eyes can't tell them apart. HD 80606, the star with the eccentric and misaligned planet, is accompanied by a nearly identical twin star located about 1,200 AU away. (You will probably not be surprised to learn that the name of the other star is HD 80607.)

I will lump together all these variations on the same theme with the name *tidally driven migration*. Take a few large bodies and allow their mutual gravitational interactions to launch a giant planet onto a highly eccentric orbit. Then, wait for tidal effects to drain energy and angular momentum away from the planet, causing it to migrate inward to a smaller and more circular orbit. Sometimes, with tongue in cheek, I call this idea the "freshman physics" hypothesis, because it's based on first-year college physics (Newton's laws of motion and gravity) supplemented with a little upper-level material about tides and friction, and because one of the two authors of a 1996 paper proposing this theory was Eric Ford, then a college freshman at the Massachusetts Institute of Technology (admittedly, not a typical freshman), working under the direction of Fred Rasio.

The tidally driven migration hypothesis has the virtue of simplicity. Of course, that doesn't mean it's correct. Many astronomers think it's too contrived to be a good explanation except in very special cases. Arranging for just the right interactions to fling a giant planet onto an appropriate orbit—close enough for tides to be important, yet not so close that the planet is destroyed—seems unlikely to happen often enough to explain all the systems we observe. An even more serious problem is that there are some giant planets whose orbits are small enough to be inside the snow line, raising the question of how they formed, and yet far enough from the star for tides to be irrelevant. Without strong tides, there can be no tidally driven migration.

Disk-Driven Migration

To set the stage for the second hypothesis, we need to visit a star called WASP-47, located 870 light-years away in the direction of the constellation Aquarius. If we could see it up close, WASP-47 would look like the Sun, at first glance. Looking more closely, we would see a hot Jupiter circling around the star every 4.2 days. Looking even more closely, we would spot two other planets, much smaller than the giant planet, with radii of 1.8 $R_\oplus$ and 3.6 $R_\oplus$. The orbits of the two little planets are aligned with that of the giant planet, with one planet on the inside and the other on the outside. Their orbital periods are 0.8 and 9 days. The discovery of the little planets was reported in 2015 by a group of astronomers led by Juliette Becker, then at the University of Michigan. They detected the planets' transits using the Kepler space telescope, a revolutionary instrument that will play a starring role in the next chapter.

It's hard to see how the hot Jupiter in the WASP-47 system could have arisen through tidally driven migration. If it was formerly a distant giant planet that got nudged or thrown onto a highly eccentric orbit, then how could the smaller planets have survived? While the giant planet was on an elliptical orbit stretching all the way from the cold environment where it formed to the immediate vicinity of the star, the planet's gravity would have had a ruinous effect on any smaller planets in the way. The small planets would have collided with the giant planet, fallen onto the star, or been ejected from the system. How could they have ended up in the delicate condition of being just inside and just outside of the giant planet's orbit?

In the frenetic year following the discovery of 51 Pegasi b, a different scenario was proposed for shrinking the orbit of a giant planet that is compatible with the observed features of WASP-47. In this scenario, all the important events take place within the first few million years after the formation of the star. If we could turn back the clock that far, we would see a giant planet forming within the protoplanetary disk, beyond the snow line. For a few million years afterward, the gaseous protoplanetary disk is still present and is available to absorb the planet's orbital energy and angular momentum.

The gravitational attraction exerted by a giant planet is not only strong enough to accrete hydrogen and helium gas from its immediate environment—it's also strong enough to pull on the gas further afield and change the way the gas is distributed in the disk. As a result, a giant planet can sculpt patterns and launch waves within the disk, within which the density of the gas is higher than in the undisturbed regions. The rearrangement of the gas within the disk causes fluctuations in the gravitational force that the disk exerts back on the planet. As a result, the planet and the disk enter into a complicated gravitational relationship. In the right conditions, the planet can transfer its orbital energy and momentum to the disk, causing the planet to spiral inward toward the star and eventually transforming it into a hot Jupiter (refer back to figure 4.5).

CHAPTER FOUR

This hypothesis is called *disk-driven migration*. The planet migrates from its initial location beyond the snow line to its much smaller final orbit by surrendering its energy and angular momentum to the protoplanetary disk. This isn't freshman physics. To calculate the patterns in the disk sculpted by the planet's gravity, and then to figure out how the rearranged gas acts back on the planet to change its orbit, requires more theoretical training than I have. These complex calculations have been performed, though, and have been validated by computer simulations that keep track of the orbit of the planet, the swirling motions of the gas within the disk, and the gravitational interactions between all the players.

The disk-driven migration hypothesis holds more promise than tidally driven migration to explain systems such as WASP-47. In this hypothesis, the initial orbits of the three planets in the WASP-47 system were all larger than they are today. After they formed, the planets slowly spiraled inward toward the star as a result of their interactions with the protoplanetary disk, never coming close enough to one another to cause mayhem. Fitting this picture, all three of the WASP-47 planets have nearly circular orbits, as would be expected from the comparatively gentle experience of slowly spiraling inward within a circular disk.

The disk-driven migration hypothesis was promulgated in 1996 by a trio of theorists led by Doug Lin, of the University of California at Santa Cruz, who were trying to explain the origin of 51 Pegasi b. The basic idea of disk-driven migration can be traced back to 1980, well before the dawn of exoplanetary science, in a paper by Peter Goldreich and Scott Tremaine, then at Caltech and the Institute for Advanced Study in Princeton, respectively. They realized that disks and planets should interact with each other and tried to calculate the effects. They showed that the giant planets of the Solar System should have undergone major changes in their orbital distances, but because they didn't have the benefit of modern knowledge about protoplanetary disks and modern computer simulations, they couldn't be sure of the details. They weren't even sure which way a planet would go: inward or outward? Their paper had no immediate applications to exoplanets in 1980, but once hot Jupiters were discovered, it became a must-read for planet formation theorists.

Since then, simulations of the gravitational interactions between planets and protoplanetary disks have been performed with increasing accuracy, and disk-driven migration seems ever more likely to occur. In fact, it seems inevitable, which has raised new problems. Can anything prevent disk-driven migration? What stopped our own Jupiter from becoming a hot Jupiter? Why would a giant planet spiral all the way inward to 0.05 AU and then stop, instead of continuing and being destroyed by the star's tidal forces?

There are no firm answers yet. The outcome of disk-driven migration probably depends on the protoplanetary disk's mass, thickness, temperature, or some combination of details that varies from star to star. One way a hot Jupiter might avoid spiraling all the way down to the stellar surface is if the inner part of the protoplanetary disk becomes ruptured by a young star's intense radiation or its strong magnetic field. That way, when the planet spirals in far enough to reach the ruptured part of the disk, there is not much gas to interact with, and the planet stops moving.

To summarize, tidally driven migration is based on the relatively simple physics of a few bodies interacting through gravity, making it relatively straightforward to perform calculations and make predictions. The randomizing effects of the gravitational interactions also have the potential to explain the eccentric and misaligned orbits that are often observed, as in the HD 80606 system. However, tidally driven migration requires seemingly contrived conditions, and cannot explain other types of systems, such as WASP-47. Disk-driven migration is based on the complex physics of a planet interacting with a vortex of gas. Computer simulations verify that the net effect is to cause planets to spiral inward. The problems with disk-driven migration are that it seems hard to stop and it has a hard time explaining eccentric and misaligned orbits.

It's not clear which of those two hypotheses is closer to the truth. Maybe both are relevant, depending on the circumstances. Or maybe the correct scenario has not yet been dreamed up. The galaxy's planetary systems are hundreds of billions of theaters showing different performances of the laws of physics. The more performances we observe, the more we will understand.

CHAPTER FIVE

THE NEW WORLDS

In the first decade of exoplanetary science, researchers were obsessed with giant planets, and had little to say about planets as small as the Earth or Venus. This obsession didn't take hold because giant planets are more interesting than small planets, but rather, because giant planets are easier to detect. The Doppler shift produced by a planet is proportional to the planet's mass. Placed in the same orbit, a Jupiter-mass planet causes its star to move about 300 times faster than an Earth-mass planet. Likewise, the transit signal produced by a planet is proportional to the planet's cross-sectional area; a Jupiter-sized planet blocks about 100 times as much light as an Earth-sized planet.

Until about 2004, almost all the known exoplanets were more massive than Saturn (95 $M_\oplus$). Yet, the more familiar we became with giant planets, the greater our hunger became to find their smaller brethren. We had so many questions. Are there small planets hugging their parent stars as tightly as hot Jupiters—and, if so, would they provide more clues about the formation of close-orbiting planets? Can small planets have eccentric and tilted orbits, like giant planets? How common are Earth-sized planets in the habitable zones of their stars? To make progress, we needed better technology.

The first explorers to reach the regime of smaller planets used the Doppler technique. One group was based at Geneva Observatory, headed by Michel Mayor, famous for having discovered 51 Pegasi b with Didier Queloz. Their fiercest competition was a group of Americans, led by Geoffrey Marcy, then at the University of California, Berkeley, and Paul Butler, of the Carnegie Institution for Science in Washington, DC. The two groups followed different paths to higher precision.

To measure a star's Doppler shift, you need a spectrograph: a device that allows you to disperse starlight into a spectrum of colors, capture an image of the spectrum with a light-sensitive detector, and measure the positions of the dark absorption lines. You need to obtain spectra night after night, to look for changing Doppler shifts consistent with the back-and-forth motion of the star around the center of mass of a planetary system. Crucially, you need to distinguish between changes in the spectrum due to Doppler shifts and changes that occur within the spectrograph in between the measurements. If the detector shifts in position relative to the mirrors and lenses, then the absorption lines in the recorded spectrum will seem to have shifted. If the detector warms up by a few degrees, the components expand slightly, causing the star's absorption lines to fall on different pixels. How can true Doppler shifts be distinguished from these other effects?

THE NEW WORLDS

The Americans relied on a trick that had been pioneered by Gordon Walker and Bruce Campbell in the 1980s. They placed a glass container of iodine gas within the spectrograph, interrupting the beam of starlight before it reached the detector. This caused the spectrum of the starlight to acquire additional dark lines due to absorption by iodine molecules. The iodine lines served as benchmarks. Any changes in the instrument would affect the star's absorption lines and the iodine lines in the same way, while any Doppler shifts in the star's lines would not be seen in the iodine lines. By focusing exclusively on the *differences* between the spectra of the starlight and of the iodine gas, their measurements became immune to many sources of error.[1] It's a beautiful example of a strategy employed in almost all ultra-precise measurements: design your experiment to be directly sensitive to small differences between similar entities, rather than trying to perform absolute measurements of a single entity.

The Geneva-led team did not follow this maxim quite as closely as the American team. Perhaps owing to the Swiss tradition of clockmaking, they relied on meticulous mechanical engineering; their strategy for minimizing the errors due to changes in the instrument was to prevent the instrument from ever changing. They built a spectrograph called the High Accuracy Radial Velocity Planet Searcher, or HARPS, which was not bolted to the back of the telescope like a normal astronomical instrument. Instead, HARPS is in a separate room and is connected to the telescope through fiber optics, allowing it

1. Paul Butler is especially proud of figuring out that iodine could be used instead of hydrogen fluoride, the gas Walker and Campbell used. Hydrogen fluoride is a corrosive acid that causes blindness and lung damage. It must have made for some tense nights at the telescope.

CHAPTER FIVE

to sit perfectly still even while the telescope is slewing around to point at different stars. HARPS has no moving parts and is encased in a sealed metal cylinder with most of the air pumped out, to minimize the variations in pressure and humidity. Inside the cylinder, the temperature is kept constant to within 0.01°C. Thanks to this extraordinary isolation and stability, HARPS can measure Doppler shifts as small as one meter per second—a walking pace.

After completing the construction of HARPS in 2003, the team began monitoring the Doppler shifts of several hundred nearby stars. For the next five years, they kept quiet about their results. Many of us began to wonder if the instrument was performing as well as had been hoped, or if their work was delayed by technical problems. Finally, in 2008, they announced their results at a conference in Cambridge, Massachusetts. They had discovered scores of planets less massive than any that had been discovered before. Some of the new planets had masses of about 20 $M_\oplus$, making them comparable to Uranus and Neptune.[2] Such planets, they claimed, exist around at least one-third of Sun-like stars, and often occur in closely spaced systems of multiple planets all having orbital periods shorter than a year. The audience was stunned. Our Milky Way galaxy was suddenly flooded with miniature solar systems. Our obsession with giant planets was about to end. The atmosphere was jubilant. When it was my turn to speak at the conference, I said it felt like performing at the Woodstock of exoplanets.

2. Technically, these were *minimum* masses and not masses, reflecting the usual limitation of the Doppler technique, but there were so many objects with low minimum masses that it was implausible for all of them to be giant planets on unfavorably oriented orbits.

Super-Earths or Mini-Neptunes?

Eventually, it became clear that the Doppler surveyors had un-veiled new types of planets that are widespread throughout the galaxy but have no representatives in the Solar System. The masses of these newly discovered planets are typically between the mass of Earth at one end and Neptune at the other end. At first, because the Doppler method specifies only a planet's minimum mass and tells us nothing about its radius, nobody knew how to interpret these new worlds. Are they extra-large solid planets, resembling scaled-up versions of the Earth? Or are they more like shrunken versions of Neptune, with thick gaseous outer layers? Put another way, if the Earth is a billiard ball, and Neptune is a basketball, we didn't know if the newly discovered balls were meant for bocce or volleyball.

As always in science, achieving a deep understanding takes a long time, but inventing new jargon is instantaneous. We began referring to the two possible kinds of planets, bocce balls and volleyballs, as super-Earths and mini-Neptunes. If we could tell the difference, we would learn whether these wide-spread planets have solid surfaces and might therefore be suit-able for life as we know it. We might also gain clues about how they formed. For example, in theory, a planet as massive as 10 $M_\oplus$ should have had a gravitational field strong enough to accrete a substantial amount of hydrogen and helium gas from the protoplanetary disk, given enough time and enough gas in the environment. Therefore, whether a 10 $M_\oplus$ planet has a substantial gaseous layer should depend on the history and lo-cation of its formation.

Most of the time, though, the only relevant clues we can obtain about an exoplanet's composition are its mass, using the Doppler method, and its radius, using the transit method, if we are lucky enough to be viewing the planetary system from a suitable direction. With the mass and radius, we can calculate the planet's overall density, its total mass divided by its total volume. I'm careful to say "overall" density because the density of the material inside a planet increases with depth. For example, the Earth's density varies from about 3 g/cm^3 in the crust to 13 g/cm^3 at the core, and its overall density is 5.5 g/cm^3.

Admittedly, the overall density isn't much information— but at least it gives theorists a starting point. They can build mathematical models of planets with different hypothetical compositions based on the physical principle of *hydrostatic balance*: at any depth within the planet, the downward force of gravity must be balanced by the upward force of pressure. The downward force is from the gravitational attraction to the interior mass, and the upward force is from the resistance to compression of the substances composing the planet. Given an assumed mixture of rock and metal, as well as lighter elements such as hydrogen, helium, and water, the principle of hydrostatic balance allows theorists to determine how the density and the pressure should increase with depth beneath the planet's surface, and predict the planet's overall density.

For example, about two-thirds of the Earth's mass is in the form of rocks (silicate minerals, with typical densities of 2–3 g/cm^3) and one-third is in the form of metals (alloys of iron and nickel, 8–9 g/cm^3). The metals, being denser than the rocks, sank to the center early in the Earth's history, when it was largely molten. There, in the core, the metals were compressed to even higher densities than usual by the weight of the overlying material. Models based on these premises can reproduce Earth's overall density of 5.5 g/cm^3.

THE NEW WORLDS

Is the same recipe for terrestrial planets used throughout the galaxy? To find out, we need to compare the predictions of planetary interior models with measurements, to see if a hypothetical composition is compatible with the data. For example, suppose we find a 10 $M_\oplus$ exoplanet. If we assume the exoplanet has the same composition as the Earth, the theoretical models predict a size of 1.8 $R_\oplus$ and an overall density of 9.4 g/cm^3. This is higher than the Earth's overall density of 5.5 g/cm^3 because the exoplanet's greater weight squeezes its own core to a higher density. If instead we assume that the exoplanet is more Neptune-ish, with 80% of the mass in the form of rock, metal, and high-pressure water, and 20% of the mass contained in an outer layer of hydrogen and helium gas, then the theoretical models predict a radius of 3.3 $R_\oplus$ and an overall density of 1.5 g/cm^3. Because of all the hydrogen and helium gas, the predicted density is lower than the Earth's density. In between these two cases is a continuum of possibilities, depending on the relative amounts of gas and heavy elements.

The planet might also have a radically different composition from any of the planets in the Solar System. A widely discussed possibility is a "water world," a solid planet in which the total mass of water nearly matches the mass of rock and metal. This is quite different from the Earth, where water covers most of the surface but constitutes only 0.02% of the planet's total mass. In another planetary system, a water world might arise if a solid planet forms beyond the snow line of a protoplanetary disk, where ice is abundant, but avoids gaining enough mass to undergo runaway gas accretion and become a gas giant. Water worlds would be planet-sized analogs of Jupiter's icy moons Ganymede and Callisto.

Given the endless possibilities for the composition of planets, it will never be possible to draw a firm conclusion about the composition of a given planet based only on the knowledge of its mass and radius. However, we can make progress by measuring the masses and radii of large samples of planets. Even if the nature of an individual planet is unknown, we might be able to match the overall trend between mass and radius with theoretical models for a single assumed composition. A large enough collection of data could reveal multiple categories of planets, each obeying a different mass-radius relationship because each has a different composition.

In 2008, seeking such trends was impossible. Because the initial discovery of the widespread existence of small exoplanets was based on the Doppler technique, and not the transit technique, we didn't know the planets' radii and couldn't calculate their overall densities. Strictly speaking, we didn't even know their masses—we knew only their minimum possible masses. The path forward was clear: we needed to discover hundreds of *transiting* planets with radii in between those of Earth and Neptune, and measure their masses. The problem was that such planets are too small for their transits to be readily detected with the telescopes then available. The effects of the Earth's atmosphere limit our ability to perform precise brightness measurements. To obtain the necessary precision, we needed a space telescope dedicated to transit-searching.

The Kepler Mission

On March 6, 2009, a Delta II rocket blasted off from Cape Canaveral, Florida, and lofted NASA's Kepler Space Telescope into its own orbit around the Sun. Once it arrived, the telescope committed itself full-time to detecting planetary transits. For four years, the telescope stared single-mindedly at several hundred thousand stars within a patch of the sky straddling the constellations of Cygnus and Lyra, with a brief interruption once a month to point its antenna toward Earth and send the data back to eager astronomers.

The first scientific publication from the Kepler mission was about a star called HAT-P-7, a star that I happened to be studying at the time using telescopes at various mountaintop observatories. The precision of my data was always frustratingly limited by the corrupting effects of Earth's atmosphere. When I first saw the graph showing the Kepler data—brightness versus time—it felt the same way it feels when I put on my glasses in the morning. The fluctuating, jagged pattern of data points to which I had become accustomed was suddenly replaced by a nearly perfect mathematical curve (figure 5.1).

The high precision of the Kepler telescope enabled the discovery of approximately 3,000 planets, enlarging the census of known planets by about a factor of five. Even more exciting was that most of Kepler's planets had sizes in between those of Earth and Neptune. Of all the transiting planets that had been detected prior to the Kepler mission, only two were smaller than Neptune, while in comparison, Kepler's haul of planets includes 2,390 that are smaller than Neptune. The Kepler mission confirmed that Sun-like stars frequently have planets larger than Earth and smaller than Neptune orbiting within 1 AU. Data from the Kepler telescope also helped to answer the question of whether the newly detected planets should be considered super-Earths or mini-Neptunes, as I'll explain later in this chapter, and led to the discovery of a variety of weird and wonderful objects, which we'll meet in the next couple of chapters.

Every astronomer I know considers the Kepler mission to have been a spectacular success. However, in a narrowly bu-

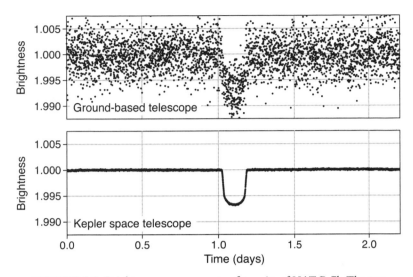

FIGURE 5.1. Brightness measurements of transits of HAT-P-7b. The top panel is based on data from the ground-based telescopes that were used to discover the planet. The bottom panel is based on data from the Kepler space telescope.

reaucratic sense, the Kepler mission failed. This is because although Kepler played a pivotal role in our understanding of super-Earths, mini-Neptunes, and other types of planets, this was not the role for which it had been designed.

The idea for the Kepler mission predated the exoplanet revolution of 1995. Throughout the 1980s, William Borucki, a scientist at NASA's Ames Research Center in Mountain View, California, dreamed of detecting exoplanets at a time when most of his colleagues considered that goal to be too farfetched to be worth pursuing. Borucki's burning question was: how common are the conditions that are conducive to life as we know it? He and his team proposed building a space telescope to perform a transit-based census of a large population of

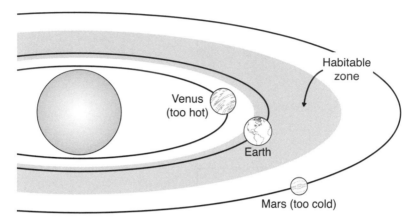

FIGURE 5.2. The habitable zone of the Solar System is calculated to lie between the orbits of Venus and Mars, with Earth near the inner edge.

Sun-like stars to determine how frequently they harbor Earth-sized planets in their habitable zones.

The habitable zone, as you may recall from chapter 1, is defined as the range of distances from a star where an Earth-like planet could plausibly have liquid water on its surface.[3] For the Sun, the habitable zone is thought to be between about 0.95 AU and 1.7 AU (figure 5.2). If the Earth were moved closer to the Sun than 0.95 AU, calculations suggest that a catastrophic greenhouse effect would ensue and cause Earth's oceans to evaporate, as is thought to have happened on Venus in the remote past. On the other hand, if Earth were placed farther than 1.7 AU from the Sun, calculations suggest it would

---

3. Because the temperature needs to be just right, neither too hot nor too cold, the habitable zone is sometimes called the "Goldilocks zone," although there is no requirement for any bears or porridge to be present.

suffer Mars's fate, becoming cold enough for all the water to freeze.

This might seem to be an arbitrary definition of "habitable." Why are we so fixated on liquid water? The answer is that liquid water appears to be crucial for life on Earth. Some living things exist without sunlight or oxygen, but there does not appear to be anything that can live indefinitely without water. Certain microorganisms and spores can survive a long time in a dried-up, dormant state, but to become active, they need water. Furthermore, it's widely accepted that life on Earth originated within liquid water, possibly in volcanic hot springs or deep-ocean hydrothermal vents.

Water isn't the only unifying principle of life on Earth. Carbon is another unifying principle; all terrestrial life is based on carbon-bearing molecules such as proteins, carbohydrates, fatty acids, and nucleic acids. The problem with using carbon as a criterion for exoplanet habitability is that carbon is everywhere. It's the fourth most common element after hydrogen, helium, and oxygen. So, while carbon might be crucial for life, requiring the presence of carbon is not a helpful search criterion; it doesn't narrow down the possibilities nearly as much as requiring liquid water, which is vanishingly rare in the context of the entire universe. Another ingredient that is common to all life on Earth is DNA. Every known terrestrial life form uses DNA as its hereditary material (except for some viruses that use DNA's close relative, RNA). To astronomers, this is fascinating but useless information, because we have no way of detecting DNA from light-years away. In contrast, we can calculate whether liquid water could plausibly exist on an exoplanet based on the characteristics of its orbit and its parent star, as well as reasonable assumptions about the planet's surface and atmospheric conditions. We can also use spectroscopy to search for water vapor in exoplanetary atmospheres, as will be described in chapter 6.

The Kepler space telescope had the world's largest digital camera (95 megapixels) of any civilian spacecraft at the time it was built. It took a long time to build. During its prolonged gestation, the French national space agency used a smaller space telescope called CoRoT to conduct the earliest space-based survey for transiting planets.[4] The CoRoT team detected a 1.7 $R_\oplus$ planet before Kepler was launched—setting the record for the smallest known transiting planet at the time—and went on to detect a total of about 30 planets. This was an appetizer before the feast of thousands of planets that Kepler would soon be serving.

Despite Kepler's unprecedented precision, in the end, the telescope was not quite capable of definitively detecting Earth-sized planets in the habitable zones of Sun-like stars. The level of fluctuations in the brightness measurements of Sun-like stars was about twice as high as had been forecasted, for reasons that are still unclear. The problem might be that the forecast assumed that all Sun-like stars are just as stable in brightness as the Sun, which only varies by about 0.001% from one day to the next. It's possible that the Sun is unusually calm and that other Sun-like stars tend to be more variable in brightness.

The problem became apparent as Kepler approached the end of its four-year mission. The Kepler team secured more funding from NASA to continue operating the telescope for another four years, hoping that additional data would allow

---

4. The French mission's original goal was unrelated to planets. Its name was CoRot, a portmanteau of Convection and internal Rotation, the two characteristics of stars that the telescope would study. After exoplanets became a hot topic, the mission design was changed, and the final "t" was promoted into a capital "T" for Transits.

CHAPTER FIVE

Earth-like planets to be detected. Soon after, though, came heartbreaking news: a critical component of the spacecraft broke. On board were four "reaction wheels," metal wheels oriented in different directions that can be spun at any desired rate up to thousands of revolutions per second. Adjusting their spin rates causes the whole spacecraft to rotate.[5] A spacecraft needs at least three wheels, one for each dimension of space, to allow it to point (and stay pointed) in any desired direction. Kepler started out with four wheels, but one of them broke in 2012. After a second wheel gave up the ghost in 2013, the telescope could no longer maintain a steady pointing, which spoiled the precision of the brightness measurements and seemed to end the mission.

NASA engineers figured out how to get around this problem and give the Kepler telescope a second life. They realized the telescope could still maintain a fixed gaze, but only for a few months at a time, and only when the telescope pointed at stars near the ecliptic. The ecliptic, you may recall from chapter 1, is the plane of the Solar System projected onto the sky, marking the paths followed by the Sun, the Moon, and all the planets. Unfortunately, the stars that Kepler had been monitoring for four years are far from the ecliptic. The search for planets would need to start from scratch, and because the telescope needed to be re-pointed every few months, there was not much hope of detecting planets with orbital periods as long as a year. Nevertheless, astronomers found many productive uses for the resurrected Kepler telescope before it ran out of fuel and shut down for good in 2018.

---

5. Yet another consequence of the conservation of angular momentum. The increased spin rate of a wheel comes at the expense of the angular momentum of the rest of the spacecraft, which rotates in response.

The top-level goal of the Kepler mission was to provide enough data to calculate the frequency of Earth-like planets, a quantity denoted by the symbol $\eta_\oplus$ and pronounced "eta-Earth." Conceptually, you can calculate $\eta_\oplus$ by counting all the Earth-sized planets that exist within the habitable zones of a large sample of stars, and then dividing by the total number of stars. For example, if every Sun-like star in the galaxy had an exact clone of the Earth, then $\eta_\oplus$ would be 1, whereas if two Earth-like planets were the norm, then $\eta_\oplus$ would be 2. If Earth-like planets only existed around one in a thousand stars, then $\eta_\oplus$ would be 0.001.

The Kepler data were not quite good enough to allow Earth-like planets to be detected, at least not with enough confidence to declare that the signal was from a planet rather than a planet-mimicking pattern of random fluctuations. Nevertheless, it's possible to estimate $\eta_\oplus$ by extrapolating from the measured frequencies of more readily detected types of planets, with larger sizes and shorter periods. Because of the need to extrapolate, the results are uncertain. In a paper published in 2021, the Kepler team stated, in effect, they would give two-to-one odds that $\eta_\oplus$ is somewhere between 0.16 and 0.85.[6] Leaving open such a wide range of possibilities might sound like a disappointing outcome from a $600 million mission. If so, keep in mind that before Kepler, any estimates of the frequency of Earth-like planets were no more than guesses. Now, we can be certain that Earth-like planets are not one-in-a-thousand. The Earth is far from unique in its size and surface temperature. With somewhat less confidence, we can

6. This range of possible values was obtained with one method of extrapolation. Using a second method, they obtained a different range, between 0.24 and 1.50, which goes to show that extrapolation is a tricky business.

CHAPTER FIVE

also rule out $\eta_\oplus > 2$. The Sun does not seem unusually impoverished in Earth-like planets.

Even if the correct value of $\eta_\oplus$ is 0.16, at the low end of the range calculated by the Kepler team, the galaxy would contain billions of potentially habitable planets. The nearest Earth-like planet would be expected to be located about 20 light-years away. Astronomically, that's just down the street from the Sun.

The TESS Mission

Even while the Kepler mission was still going strong, NASA was investing in the construction of Kepler's successor, which is called the Transiting Exoplanet Survey Satellite, or TESS. The visionary in this case was George Ricker, of the Massachusetts Institute of Technology, and I'm fortunate to be on the team that has been working under his direction since 2006. We didn't design TESS to search for Earth-like planets. By 2006, we were aware of hot Jupiters and other giant exoplanets, and as the concept of our mission matured, we learned about the plethora of planets in between the Earth and Neptune in size. We knew that these smallish planets posed many interesting questions and would be easier to detect and study than habitable-zone Earth-sized planets. We designed TESS to discover hundreds of transiting planets smaller than Neptune around the nearest and brightest stars in the sky.

Brightness is of paramount importance in astronomy. All other things being equal, if a star is brighter (because it is closer to us, or intrinsically more luminous, or both), we can extract more information from its light. When we perform Doppler spectroscopy, for example, we need to measure the intensity of the starlight at a hundred thousand different wavelengths. If the starlight is feeble to begin with, there's no hope of measuring the intensity after the light's total energy has been divided into such small slices.

Most of the stars that the Kepler telescope observed are annoyingly faint. The reason is that the Kepler telescope's field of view spanned only one four-hundredth of the sky, similar in size to the area of the sky you can cover with a fist at arm's length. The brightest stars are sprinkled all over the sky, with most of them outside of Kepler's field of view. The only way to arrange for Kepler to monitor several hundred thousand stars was to include the multitudes of faint stars that happened to appear within the telescope's field of view. Most of those stars are thousands of light-years away. The consequence is that most of the planets that Kepler discovered are orbiting stars that are too faint for Doppler spectroscopy, which, in turn, means we cannot measure the planets' masses.

TESS overcomes this problem by scanning the entire sky, making it possible to observe all the brightest stars. Instead of one telescope, the TESS spacecraft has four telescopes that point in different directions. The telescopes' combined field of view spans an angular area with dimensions 24° by 96°, about 20 times larger than the Kepler telescope's field of view. TESS was launched into space from Cape Canaveral by a SpaceX rocket on April 18, 2018. For the initial TESS survey, the entire sky was divided into 26 sectors. TESS stared at each sector for one month before moving to the adjacent sector for another month, and so forth. Over the course of two years, TESS worked its way around most of the celestial sphere (plate 8). It then began a series of three-year surveys to reobserve sectors while also filling in the remaining gaps in sky coverage.

CHAPTER FIVE

Using this strategy, a typical star is observed by TESS for one month at a time, with a few years in between these one-month visits. For the purpose of finding planets in the habitable zones of Sun-like stars, this strategy is not ideal. Such planets have orbital periods close to a year. Because the transit probability is so low, and because we need to observe at least three transits for a secure detection, we'll need to wait until TESS has gathered at least three years' worth of data for thousands of bright stars before we can expect to find any Earth-like planets. In the meantime, TESS is picking the lowest-hanging fruit from the exoplanetary orchard: planets with periods shorter than a few weeks. The short-period planets are providing plenty of opportunities for interesting discoveries while we gradually accumulate more data and extend our reach to longer-period planets.

An important difference between Kepler and TESS is that the TESS spacecraft is much closer to the Earth. Kepler was launched into its own orbit around the Sun, which was chosen to be a little wider than the Earth's orbit. So, by virtue of the third law of the Kepler telescope's namesake, Johannes Kepler, the spacecraft had a longer orbital period than Earth. The difference in orbital periods caused the telescope to lag behind the Earth and grow increasingly distant. The advantage of this orbit was that the Kepler telescope could operate in total darkness and seclusion, with no stray light from the Earth or Moon. The disadvantage was that phoning home was difficult. The long distance limited the amount of data that could be sent back to Earth. By the time the mission ended, Kepler was 0.92 AU away from the Earth. In contrast, TESS is a homebody, never straying from the Earth by more than 0.003 AU. TESS's orbit is an ellipse that loops around the Earth with a period of 13.5 days. Most of the time, TESS is far enough from Earth to enjoy dark skies, and when TESS does approach the Earth at the fast end of its elliptical orbit, the spacecraft is close enough to transmit 100 gigabytes of data from its onboard computer. We receive much more data from TESS than we did from Kepler, allowing TESS to search for planets around many more stars than Kepler, while also providing a voluminous dataset for other types of astronomical objects, such as asteroids, pulsating stars, and supernova explosions.

We hope to keep re-scanning the sky with TESS for as long as the spacecraft stays in good health. By 2022, TESS identified about 6,000 stars with periodic brightness dips, most of which are probably due to transiting planets, although many are still waiting to be confirmed by collecting more TESS data and by performing complementary observations with other telescopes. Eventually, we expect TESS will discover even more planets than the Kepler mission—and the TESS planets will be closer to home, orbiting the nearest and brightest Sun-like stars in the galaxy.

Four Plots from the Frontier

You are now fully equipped to visit the frontier of exoplanet discoveries with the Doppler and transit techniques. To summarize our progress and the boundaries of our knowledge, I'd like to show you a series of four plots. They are well worth the effort to understand.

In the first plot (figure 5.3), each dot represents a planet that was detected using the Doppler method. With this method, the two things about an exoplanet that we can measure best are its orbital period, conveyed by the dot's horizontal posi-

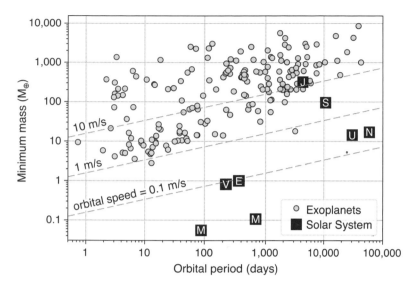

FIGURE 5.3. Planets in the California Legacy Survey. Shown for each planet are the minimum mass and orbital period. The upper left part of the plot is where planets are easiest to detect. The lower right part is where planets may exist but cannot be detected because they have low masses or long periods. For comparison, the black squares represent Solar System planets.

tion, and its minimum possible mass, conveyed by its vertical position. The dark squares represent planets in the Solar System. It might seem odd that the tick marks on the axes are not evenly spaced. The small tick marks get closer together and bunch up as they approach one of the bigger tick marks. That's because this is a *logarithmic* plot. In an ordinary plot, a single scale factor is chosen for each dimension; for example, if you move one centimeter to the right, the orbital period goes up by one day. In astronomy, ordinary plots are often ineffectual. We deal with objects having properties that range over many orders of magnitude—in this case, planets with orbital periods

THE NEW WORLDS

between 1 and 100,000 days. With an ordinary plot, if we were to extend the horizontal axis to 100,000 days to encompass the longest-period planets, then all the planets with periods between 1 and 1,000 days would be crammed together close to the vertical axis, and we wouldn't be able to make out any details. On a logarithmic plot, the scale factor grows as we move from one end of the axis to the other end. The first vertical grid line corresponds to one day, the second one to 10 days, then 100 days, 1,000 days, and so forth.[7] By using a logarithmic plot, we can obtain a clear view of all the data points on a single set of axes.

Rather than including all the Doppler planets ever discovered, figure 5.3 is based on the results of a single comprehensive study, the California Legacy Survey, undertaken by a group of astronomers led by Andrew Howard of the California Institute of Technology. By focusing on a single survey, we avoid the confusion that arises when trying to interpret the combined results of different surveys with different levels of sensitivity to planets. The California Legacy Survey was based on monitoring the Doppler shifts of 719 stars over a time interval spanning 30 years. The earliest measurements were obtained by the Doppler pioneers, predating the discovery of 51 Pegasi b. In total, the survey consists of 100,000 individual measurements of Doppler velocities, a feat that will not soon be surpassed. They detected 177 planets, ranging in minimum mass from 3 $M_\oplus$ to 7,000 $M_\oplus$ and in orbital period from 18 hours to 100 years. The longest-period planets were especially diffi-

7. Mathematically, with every step on the chart, the logarithm of the scale factor increases by one unit. The logarithms (in base 10) of the numbers 1, 10, 100, and 1,000 are 0, 1, 2, and 3.

cult to detect because they did not even complete a single orbit during the survey.

The upper left part of figure 5.3 is filled with data points, but there appears to be a diagonal boundary beneath which there are very few. The approximate boundary is marked by a dashed line labeled "1 m/s." Does the absence of dots below that dashed line, in the lower right part of the diagram, imply that planets with long periods and low masses are rare? Definitely not. The absence of planets in the seemingly forbidden region is due to the limitations of the survey; such planets are too difficult to detect. Any planet with properties that land on the "1 m/s" dashed line would cause its star to wobble with a speed of about one meter per second, which is about the smallest signal that can be reliably detected for a typical star. Two other dashed lines are drawn, to show where the maximum Doppler speed would be 0.1 m/s and 10 m/s.

Even at the top of the plot, where the giant planets are on display, our view is distorted by technological limitations. Short-period planets are easier to detect than long-period planets, causing the survey to be biased. For example, if we find comparable numbers of short-period and long-period planets of a given mass, despite the greater difficulty of detecting long-period planets, then long-period planets must be intrinsically more common than short-period planets. After correcting for survey bias, the California team concluded that for every 100 Sun-like stars in the galaxy, there are approximately 6 giant planets with orbital distances smaller than 0.75 AU (about the size of Venus's orbit), and 25 giant planets with orbital distances between 0.75 and 30 AU (between Venus's and Neptune's orbital distances). In this study, a giant planet was defined as a planet with a minimum mass between 30 $M_\oplus$

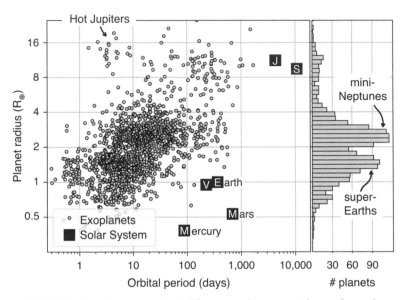

FIGURE 5.4. Planets in the California Kepler survey. Shown for each planet are the radius and orbital period. For comparison, the black squares represent Mercury, Venus, Earth, Mars, Jupiter, and Saturn. On the right side is a histogram of planet sizes, which shows two peaks, corresponding to super-Earths and mini-Neptunes.

and 6,000 $M_\oplus$, or equivalently, about 0.1 to 20 times the mass of Jupiter.

The second plot (figure 5.4) is another scorecard from an exoplanet survey, but in this case, it's a transit survey rather than a Doppler survey. The vertical position of each dot conveys the planet's radius, not its mass. I chose to plot the data from the Kepler survey[8] rather than the newer TESS survey, because at present the Kepler survey is more comprehensive. If we re-

---

8. To be specific, it's a subset of the Kepler survey called the California Kepler Survey, for which the planet radii were measured in a consistent manner.

strict our attention to the left side of the plot, where the orbital periods are shorter than about 100 days, we see many more small planets (1–4 $R_\oplus$) than giant planets (4–16 $R_\oplus$), even though small planets are harder to detect. Ergo, small planets are intrinsically more abundant than giant planets within the inner regions of planetary systems. The misplaced giants—those with sizes comparable to Jupiter and orbital periods shorter than a year—are fascinating, and demand revisions to our theories of planetary formation and evolution, but they are revealed here to be a small minority.

The transit plot, like the Doppler plot, has boundaries outside of which there are no detected planets. There are very few planets on the far-right side of figure 5.4 because the Kepler mission did not last long enough to probe for planets with longer periods. And there are very few planets below the diagonal dashed line, because the random fluctuations in the brightness measurements prevented the detection of such small planets.

The right side of figure 5.4 depicts a histogram (plotted sideways) that shows the relative abundances of planets of different sizes. There are two peaks in the histogram, implying that planets come in two especially common sizes. The two peaks are centered at 1.4 $R_\oplus$ and 2.4 $R_\oplus$. There are relatively few planets in between these two peaks, a feature of the data that has been called the "radius valley." The two peaks are not due to bias or any other artifact of the survey. They are genuine evidence for two distinct categories of planets.

The existence of the two peaks separated by the radius valley answers the question I posed earlier: should the small planets that started turning up in 2008 be considered super-Earths or mini-Neptunes? The answer is: yes. There appear to be two different recipes to make planets with sizes in between those of Earth and Neptune. The planets with radii near 1.4 $R_\oplus$ are probably balls of rock and metal, similar in structure to the Solar System's terrestrial planets, but more massive and therefore deserving of the name super-Earths. Those with sizes near 2.4 $R_\oplus$ are thought to be balls of rock and metal that are enshrouded by hydrogen and helium gas. The typical mass of the gas layer is probably only a few percent of the planet's total mass, but because hydrogen and helium are so lightweight, the gas causes the planet's outer radius to expand to as much as twice the size of the underlying solid part of the planet, forming a mini-Neptune.

The main evidence for these claims about the differing structures of the newly discovered planets comes from the combination of the Doppler and transit techniques. For those planets where both techniques have been applied, allowing us to calculate the planet's overall density, the super-Earths have high densities consistent with rock and metal, and the mini-Neptunes have lower densities that seem to require lightweight gases in the mixture. There are certainly other possibilities for the structures of these planets—as I emphasized earlier, the overall density is only one number, which does not suffice to infer a planet's internal structure and composition. So, we shouldn't take the names "super-Earths" and "mini-Neptunes" too seriously, particularly because we lack hard knowledge about the interior of Neptune itself. However, there is a point in favor of the super-Earth / mini-Neptune interpretation that impresses me. The radius valley was *predicted* by theorists. Theory failed to predict some of the key discoveries in exoplanetary science, but in this case, theorists got there first. Two studied published in 2013, one led by James Owen (then at the Canadian Institute for Theoretical Astrophysics in Toronto) and the other by Eric Lopez (then at the University of

California, Santa Cruz) predicted that planets with sizes of about 2 $R_\oplus$ would be rare in comparison to smaller or larger planets. Their theory invokes a remarkable planet-wide transformation: a mini-Neptune that blows off its outer gaseous layer to reveal a super-Earth underneath.

The theory was developed after we knew that Earth-to-Neptune–sized planets are commonly found with orbits smaller than 0.5 AU, and after we knew that some of those planets had low densities. The low densities were a clue that these planets accreted hydrogen and helium gas from the protoplanetary disk. Standard planet-formation theory held that this was possible only for planets located beyond the snow line, but Owen and Lopez proposed that the low-density planets formed close to the star, perhaps even right where we see them today. In their scenario, the planets formed as solid bodies with a typical mass of about 5 $M_\oplus$, just high enough to accrete a little bit of gas from the protoplanetary disk, but not high enough to undergo runaway gas accretion and become giant planets. Because they had lower masses than giant planets, and weaker gravity, their hydrogen–helium atmospheres were not held as tightly. This made the atmospheres vulnerable to being lost. During the first hundred million years after their formation, stars are prone to emitting bursts of high-energy radiation such as ultraviolet rays and X-rays. Any planet that forms too close to such an unruly star is in danger of having its atmosphere heated by high-energy radiation, causing the atmosphere to expand to such a degree that it escapes the planet. No longer gravitationally bound to the planet, the gaseous atmosphere flows away, exposing the planet's solid interior. Crucially, according to the theorists' calculations, losing the hydrogen–helium atmosphere is an all-or-nothing process. Either the planet retains enough gas to maintain a size

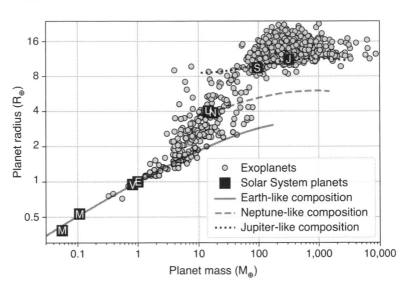

FIGURE 5.5. Radius versus mass for transiting planets, along with Solar System planets for comparison. The three curves are theoretical relationships between radius and mass for three different assumptions about a planet's composition.

exceeding 2 $R_\oplus$, or the entire inventory of gas is lost, leaving a naked ball of rock and iron. That could be why we find mini-Neptunes and super-Earths, with nary in between.

Turn now to the third plot (figure 5.5), which shows the masses and radii of all the planets for which both quantities have been reliably measured. Each dot represents a planet, and the three curves are theoretical predictions based on different assumptions about a planet's composition. For example, the solid curve represents an Earth-like composition: a rocky mantle with two-thirds of the total mass, and an iron-nickel core containing the other one-third of the mass. The dashed curve represents planets with a similar composition to those of Uranus and Neptune, which are thought to consist of 80–90%

heavy elements and 10–20% hydrogen and helium. The dotted curve is for planets composed almost entirely of hydrogen and helium, such as Saturn and Jupiter.

Start your tour of figure 5.5 on the lower left side, with the smallest planets. Here we see what I previously asserted: the data for almost all the planets smaller than about 2 $R_\oplus$ are consistent with an Earth-like composition. The next higher group of planets on the chart, between 2 $R_\oplus$ and 4 $R_\oplus$, tells a different story. Their overall densities are too low to be compatible with an Earth-like composition, but there is no single theoretical curve that runs through all the data points. Evidently, the planets in this size range have a variety of different compositions, perhaps differing in the amount of hydrogen-helium gas that they were able to accrete, or in the abundance of relatively lightweight molecules such as water, methane, and ammonia.

Continuing further upward, we encounter a relatively empty part of the chart between 4 $R_\oplus$ and 8 $R_\oplus$ before we enter the realm of the giant planets, between 8 $R_\oplus$ and 18 $R_\oplus$. The giant planets, too, tell an interesting story. Some of the data points lie far above the dotted curve representing a Jupiter-like composition—that's the "radius inflation problem" that was described in chapter 4. We don't yet know why these giant planets are so swollen.

There's another pattern in the data that I find fascinating. Taken together, the giant planets trace out a nearly horizontal thick stripe. The collection of planets with masses of about 100 $M_\oplus$ (approximately Saturn's mass) have nearly the same average radius as the planets with masses of 10,000 $M_\oplus$ (about thirty times more massive than Jupiter). The dotted curve, representing the theoretical prediction, shows this behavior, too: to the right of Jupiter, the curve levels off to become a nearly horizontal line. I hope you'll agree that this is strange. If we increase the mass of an object, we expect the object to become larger than when we started. Giant planets aren't like that. Instead, they stay nearly the same size and become denser.

The reason for this counterintuitive behavior is that a sufficiently massive planet starts to collapse under its own weight. Ordinary materials resist being compressed; it would be an amazing feat of strength to squeeze a billiard ball down to the size of a golf ball. Even hydrogen and helium, usually imagined to be wispy gases, will fiercely resist compression if they are confined to a small enough volume. However, near the center of a giant planet, the weight of all the overlying material is high enough to overwhelm even the most rigid materials. The resulting compression causes the planet's overall density to increase.

If the material's rigidity gives way, what prevents the planet from collapsing all the way down to a single point and becoming a black hole? What other forces can the planet call upon to oppose the force of gravity? For these questions, unlike many others in exoplanetary science, we already have a theory that provides good answers. It involves the behavior of electrons—so, please buckle up for a discussion of quantum mechanics.

The interior of a planet more massive than Jupiter is compressed to such a high density that it's no longer gaseous, and it's not liquid or solid, either. The atoms and their electrons are squeezed so tightly together that they risk violating a quantum-mechanical rule called the *Pauli exclusion principle*. This principle, taught in introductory chemistry classes, states that no two electrons in the same system can have exactly the same properties. Unlike some other fundamental particles, two electrons are never found together in the same location, with the same energy, the same orbital angular momentum, and the same spin angular momentum. This principle helps to explain the structure of the periodic table of the elements. To add an electron to an atom, the electron must have different properties from all the other electrons in the atom, either by spinning in a different direction, or by occupying a wider orbit around the nucleus. This is what gives each type of atom a different arrangement of electrons and unique predilections for undergoing chemical reactions.

Inside a sufficiently massive planet, the electrons are pressed together as closely as the Pauli principle allows, forming a quantum-mechanical substance called a *degenerate gas*.[9] Since pressing the electrons closer together is against the rules, loosely speaking, the electrons begin to feel a repulsive force that resists further compression. This outward *degeneracy pressure* rises in strength with increasing density, which is what allows it to serve as a "pressure of last resort" and protect an object from gravitational collapse.

When more mass is added to an object supported purely by ordinary material forces, the object grows in size, at least a little, conforming with intuition. In contrast, adding mass to an object supported purely by degeneracy pressure causes the object to shrink. The object's increased weight exerts greater pressure on the interior and causes it to contract, which in turn causes the degeneracy pressure to rise. The inward and outward forces come back into balance only after the object shrinks and becomes denser. (If the added mass becomes too large, then the object shrinks all the way down to become a

9. The word *degenerate* always strikes me as unnecessarily pejorative. The gas is not deteriorating or immoral. In this context, *degenerate* is a mathematical term referring to a special case, often a case in which things that are usually distinct become the same.

neutron star or black hole, but let's save that story for chapter 7.) Giant planets are partially supported by material strength and partly by degeneracy pressure. Because of this mixed pressure support, adding mass to a giant planet causes the planet neither to grow nor shrink. The radius stays basically the same, even if its mass is increased by a factor of a thousand.

Is there a maximum mass for a planet? Once a planet's mass is increased all the way to 25,000 $M_\oplus$ (80 Jupiter masses), just beyond the rightmost limit in the mass–radius plot, the planet's core becomes hot and dense enough to ignite nuclear fusion reactions, which convert hydrogen into helium and release lots of energy. The planet becomes a star. At some point along the way between planets and stars, astronomers stop calling the object a planet and start calling it a brown dwarf, although there is no consensus about where to draw the dividing line. Some astronomers advocate an upper limit of 4,000 $M_\oplus$ (13 Jupiter masses), the point at which the core is hot and dense enough for nuclear fusion reactions involving deuterium, a rare isotope of hydrogen that is more easily ignited than ordinary hydrogen. However, fusing deuterium doesn't release much energy, and hardly changes the outward appearance of the object, making the deuterium-burning limit seem arbitrary. I think we will simply need to suppress our academic compulsion to classify things into discrete categories and live with the fact that Mother Nature does not always feel the same urge.

The fourth and final chart in the sequence (figure 5.6) displays one of the most extraordinary findings of exoplanetary science: miniature systems of super-Earths and mini-Neptunes, with all the planets packed tightly together around the same star. The chart displays the systems known to have at least five transiting planets. Each horizontal line represents a single

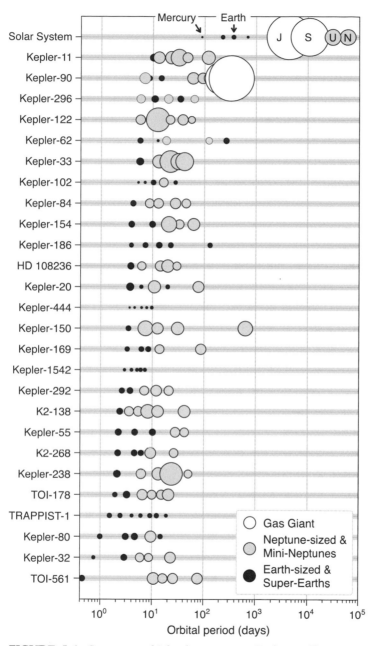

FIGURE 5.6. Compact multiple-planet systems. Each row illustrates a planetary system with at least five planets. The horizontal axis shows the orbital period, and the circles indicate the planets' relative sizes. The systems are ordered according to the shortest period. The top row is for the Solar System.

134

planetary system, and each circle represents a planet. The size of the circle is proportional to the planet's radius, the shading specifies the size category to which the planet belongs, and the horizontal position conveys the planet's orbital period on a logarithmic scale. The topmost line is for the Solar System. The exoplanetary systems consist mainly of planets with sizes in between those of Earth and Neptune. Most of the exoplanets have orbital periods shorter than Mercury's 88-day period.

One of the earliest discoveries of such a compact multiple-planet system was Kepler-11, represented by the second line from the top. This system has five known planets with orbital periods shorter than 120 days, implying that all the orbits are smaller than Venus's orbit around the Sun (as depicted in figure 5.7). One of the most striking systems is Kepler-1542 (a little more than halfway down in figure 5.6), which has a set of Earth-sized planets crowded together like adjacent cars on a five-lane racetrack, with orbital periods of approximately 3, 4, 5, 6, and 7 days. Interplanetary exploration would be a lot easier for any inhabitants of those planets than it is within our more spread-out Solar System.

A subtle feature of the multiple-planet systems shown in figure 5.6 is that the planets within a given system tend to have similar sizes. If you compare any two super-Earths orbiting the same star, their sizes tend to be more alike than if you had randomly drawn two super-Earths from the entire collection of planetary systems. The same is true of mini-Neptunes. In addition, the circles representing the planets within a single system tend to be evenly spaced in the horizontal direction. Because the horizontal axis is logarithmic, the regular spacing implies that the orbital periods tend to follow a nearly geometric progression—as you move outward, the period of each

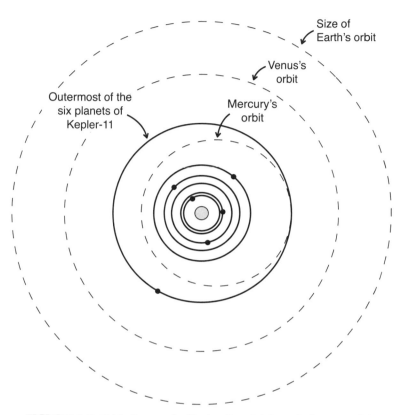

FIGURE 5.7. Orbit diagram for Kepler-11, a tightly packed system of six planets. For comparison, the dashed circles show the orbital distances of Mercury, Venus, and Earth around the Sun.

orbit is a certain multiple of the period of the previous orbit. Their regular spacings and similar sizes cause planets to resemble "peas in a pod," in a memorable analogy expressed by Lauren Weiss, of the University of Notre Dame, who led a statistical study of this phenomenon. We don't yet have a detailed physical explanation for why planetary systems resemble peapods, although it does make sense that planets that form

THE NEW WORLDS

within the same region of a single protoplanetary disk should resemble one another, just as siblings in the same family are more alike than randomly chosen people of the same ages.

Statistical analyses of the Kepler survey indicate that somewhere between 30% and 60% of Sun-like stars have a system of three or more planets larger than the Earth, all orbiting within 1 AU. The high prevalence of these compact planetary systems was a surprise; they were not predicted by planet formation theory. I would love to see how a protoplanetary disk manages to give birth to a system like Kepler-11 or Kepler-1542. Do the planets form in the compact configurations in which we observe them today? Based on what we thought we knew about protoplanetary disks, the total mass of the planets in those and similar systems is too large to have plausibly been drawn from such a small area of the disk. We could be wrong, though. Maybe disks have more solid material than we thought, or maybe the solid particles can be concentrated by aerodynamic forces more efficiently than we expected.

It's also possible the planets formed in more widely spaced orbits, and that disk-driven migration (described in chapter 4) is responsible for shrinking the orbits and bringing all the planets close together. But how could all the planets spiral inward at the same time and avoid colliding? Some of the systems appear to be right on the edge of stability. For example, if we were to double the masses of all the Kepler-11 planets, the system would not last for long. Because of the increased gravitational forces, the planets would pull four times as hard on each other. Computer simulations show that the planets would alter each other's orbits by enough to cause at least some of them to collide, fall into the star, or escape the star's gravitational pull and fly away.

CHAPTER FIVE

Before closing this chapter, I want to direct your attention back to figures 5.2 and 5.3, which show the results of a Doppler survey and a transit survey. The black squares represent the planets of the Solar System, as they would appear on analogous plots made by alien astronomers searching for planets around the Sun. It's sobering to see that all the planets of the Solar System would be difficult to detect with our current technology. Mercury and Mars are too small. The Earth and Venus would be invisible in Doppler surveys and only barely detectable in a Kepler-like survey. Jupiter (and maybe Saturn) could be detected with the Doppler method, but because of their wide orbits, transits of such planets are unlikely to occur. Any direct analogs of Uranus and Neptune are well outside the range of detectability.

Sometimes, newspaper articles and science magazines state that exoplanetary science has revealed the Solar System to be anomalous—that what we once thought were "strange new worlds" are actually the norm. I don't think we can draw this conclusion, at least not yet. We do know that roughly half of Sun-like stars have planetary systems with properties quite different from those of the Solar System, with either a misplaced giant or a tightly packed collection of super-Earths and mini-Neptunes. As for the other half of the stars, for which our current methods don't reveal any planets, we need to reserve judgment. Maybe most of those stars have planets that do resemble our friends in the Solar System, and we will simply need to work harder to find them.

CHAPTER SIX

STRANGE NEW WORLDS

The success of the transit method transformed exoplanetary science not only by revealing new types of planets, but also by attracting people to the field who think differently: planetary scientists. Until then, the field was mainly the province of astronomers and astrophysicists. You might find it surprising that planetary scientists were not on board from the very beginning, or even that there is any distinction between a planetary scientist and an astronomer. For that matter, what's the difference between an astronomer and an astrophysicist?

Astronomy, the careful observation of celestial phenomena, is a cultural activity dating back to the earliest known civilizations. Astrophysics, the effort to understand celestial phenomena in terms of universal laws of physics, began with Isaac Newton in the seventeenth century, although Newton didn't call himself an astrophysicist. The word "astrophysics" began to enjoy wide usage in the late nineteenth century with the advent of spectroscopy, which was applicable to both the light emitted by stars and the light emitted by flames and hot gases in a laboratory, thereby bringing astronomy and experimental physics into a closer union. These days, the terms "astronomer" and "astrophysicist" are nearly interchangeable, with "astronomer" being the friendlier term and "astrophysicist" being preferred when one is trying to impress somebody.[1] Planetary science began as a subfield of astronomy, but in the 1950s and the subsequent Apollo age, planetary science metamorphosed into a more interdisciplinary field. Once it became possible to send spacecraft to the Moon and the other planets in the Solar System, our knowledge of the atmospheres, surfaces, and interiors of those bodies grew exponentially. As a result, modern planetary science draws just as much from geology, geochemistry, atmospheric physics, and the other Earth sciences as it does from astronomy.

Astronomers and planetary scientists have different expectations for the amount of data they can obtain. An astronomer might be satisfied to publish a paper reporting measurements of the mass and radius of an exoplanet, knowing how much painstaking work is required to obtain that information. A planetary scientist would be underwhelmed by a dataset consisting of only two numbers. Planetary scientists want answers to questions like these: how does the temperature and composition of the planet's atmosphere change with altitude? Does the planet have continents, plate tectonics, oceans, and global

---

1. On the other hand, when one is trying to end a conversation, it's best to say you are a "physicist."

climate cycles? Which minerals are most abundant in the planet's crust, mantle, and core? How strong is the planet's magnetic field? Does it have rings or moons?

It will be a while before exoplanet astronomers can offer satisfying answers to those questions. Nevertheless, planetary scientists began talking to astronomers more frequently after the success of the transit method. They were attracted by the newly discovered categories of planets and the wide range of possibilities for their atmospheres, interiors, and orbital arrangements. They also appreciated that even if astronomers can measure only a few numbers per planet, once you have thousands of planets, you're starting to talk about real data.

The flow of planetary scientists increased further when astronomers developed techniques to measure subtler properties of exoplanets—beyond mass and radius—with a level of precision I never would have expected to be possible. This chapter is about those more detailed types of observations, and the exotic types of planetary systems they have revealed.

As usual in astrophysics, progress was made possible by developing new ways to extract information from starlight, and by applying our knowledge of physics to interpret the data. The examples in this chapter fall into two categories. In the first category, we use our knowledge of *gravitational dynamics*—the motion of bodies under the influence of their mutual gravitational forces—to deduce the three-dimensional orbital architecture of planetary systems. Considerations of gravitational dynamics have led to new clues about the formation of planetary systems as well as new mysteries. In the second category, we use our knowledge of *spectroscopy*—the study of the intensity of a light source as a function of wavelength—to study the constituents of an exoplanet's atmosphere.

CHAPTER SIX

Gravitational Dynamics

When Newton explained Kepler's three laws of planetary motion as consequences of more fundamental laws of gravity and motion, he realized that Kepler's laws were only an approximation to reality. Strictly speaking, the equations Newton solved to show that a planet travels in a perfect ellipse are only valid if the star and the planet are the only two bodies in the system. Newton and his successors knew that for a truly accurate accounting of the observed motion of the planets, they needed to include the effects of gravitational forces exerted on each planet by all the other planets, not just by the Sun. The forces between the planets cause their orbits to deviate from perfectly elliptical trajectories in complicated ways, making it difficult to predict their locations in detail.

Solving Newton's equations of motion when there are more than two gravitationally interacting bodies is an exercise that exhausted many of the best brains of the eighteenth and nineteenth centuries. If the two-body problem is a walk in the park, the three-body problem is a slog in the jungle, to say nothing of four or more bodies. After becoming aware of the problem's complexity, Newton worried that if the Solar System strictly obeyed his equations of motion, it would be unstable. Eventually, planets would collide, planets would fall into the Sun, or planets would be ejected. It was hard to believe that *any* mathematical solution of the tangled equations would exhibit nearly circular and well-aligned orbits that persisted for eons. In some of his writings, Newton speculated that divine intervention was required from time to time—a strategically timed push on Saturn, a tug on Jupiter—to keep everything tidy and ward off catastrophe. In the words of historian Michael Hoskin, "His God demonstrated his continuing concern for his clockwork universe by entering into what we might describe as a permanent servicing contract." I wonder how Newton would have reacted to the news of billions of other planetary systems for which service calls would be much more frequent. As we'll see, the interplanetary forces in some of the newly discovered planetary systems are orders of magnitude stronger than they are in the Solar System.

Today, we can accurately solve the problem of N gravitationally interacting bodies, where N is a number that can be as high as 10 million, or even 10 billion under certain approximations. We're not any smarter than Newton, but we can perform computer simulations. In a simulation of a planetary system, we choose initial locations, speeds, and directions for all the planets. The computer calculates the forces on each planet from the star and from all the other planets, allows each planet to move a little bit in response to the forces, then recalculates the forces with the star and planets in their new locations, and so forth. These simulations can predict the future of a given planetary system and expose any interesting and potentially observable consequences of interplanetary forces.

In the Kepler-11 system, for example, there are six transiting planets all within 0.5 AU of the star. In the absence of interplanetary forces, the transits would repeat like clockwork, with the same amount of time elapsing from one transit to the next. Things are a little different when we consider the gravitational forces between planets. Every time two of the planets approach each other closely, they pull hard on each other, changing their orbital speeds and causing their transits to occur on a more complicated schedule. The transits of planet Kepler-11f occur as much as half an hour earlier, or later, relative to the time that one would predict without any interplanetary forces. The exact transit times depend on the recent history of the interactions between planets.

The importance of *transit-timing variations* for exoplanetary science was predicted in 2005, prior to any such observations, and should therefore be scored as another point for theorists. At that time, nobody knew about compact systems such as Kepler-11, but two groups of theorists, led by Matthew Holman of the Smithsonian Astrophysical Observatory and Eric Agol of the University of Washington, worked out some hypothetical cases. Suppose alien astronomers were observing the Sun from light-years away and managed to detect transits of Venus. They would notice that the transits recur every 224.701 days on average, but with variations as large as 0.007 days (10 minutes). Those variations are mainly due to Venus's gravitational attraction to the Earth.

Holman, Agol, and others realized that transit-timing variations would give astronomers a sneaky way to determine the masses of exoplanets that avoids the more arduous task of tracking the star's Doppler shift. The strengths of interplanetary forces are proportional to the product of the planets' masses. By performing many computer simulations of a planetary system with different choices for the planets' masses, it is sometimes possible to determine the masses by requiring the simulations to agree with the transit-timing data. It's even possible to discover a new planet in the system by observing the irregular variations it causes in the transit times of a known planet.

Discovering a new planet by observing its gravitational effects on a known planet has an illustrious history. Uranus and Neptune were unknown to ancient astronomers because they are both too faint to be seen by eye. Telescopes are required. Uranus was discovered by accident in 1781 by William

Herschel, who had been scanning the sky with his telescope for other reasons. In contrast, Neptune's existence was *predicted*, using pen, paper, and the human mind. Urbain Jean Joseph Le Verrier, a specialist in gravitational dynamics at the Paris Observatory, was trying to explain the irregularities that had been observed in Uranus's orbit as the gravitational effects of a hitherto unknown planet. After laborious mathematical calculations, in 1846, he wrote to Johann Gottfried Galle at the Berlin Observatory to tell him where to point his telescope if he wanted to discover a new planet. Galle took the prediction seriously. Soon after he received Le Verrier's letter, he and his research assistant Heinrich Louis d'Arrest spotted Neptune within only one degree of the predicted location.[2]

One hundred and sixty-six years later, another specialist in gravitational dynamics achieved a similar feat for an exoplanetary system. In 2012, a team led by David Nesvorny, at the Southwest Research Institute in Boulder, Colorado, discovered a giant planet by timing the transits of a previously known planet orbiting the same star. The newly detected planet, Kepler-46c, does not transit the star, but the irregular transit schedule of planet Kepler-46b (figure 6.1) provided enough information for Nesvorny's team to calculate the orbital characteristics of Kepler-46c and specify its mass with a precision of 5%. To complete the analogy to the discovery of Neptune, we would need a modern-day Galle to confirm the existence of the new planet by seeing it through a telescope. Unfortunately, at 2,600 light-years away, the star is much too far away for the direct-imaging method to be feasible. The star is also

2. Another brilliant mathematician, John Couch Adams, predicted the existence of Neptune independently of Le Verrier, although he did not have as much success in convincing astronomers to act quickly on his prediction.

CHAPTER SIX

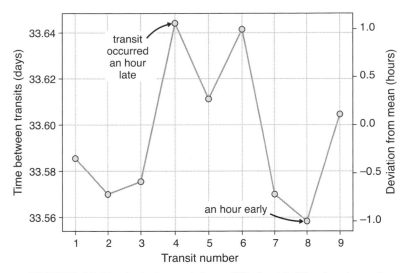

FIGURE 6.1. Transit-timing variations of Kepler-46b. The time interval between transits is not always the same, because of the gravitational forces exerted by a neighboring planet.

too faint to detect Kepler-46c with the Doppler method. If we are patient, though, it will eventually be possible to confirm the discovery using the transit method. Although Kepler-46c isn't currently a transiting planet, Nesvorny and his team predict that it will *become* a transiting planet by around the year 2035, because its orbit is gradually being reoriented by the gravitational effects of Kepler-46b. In the meantime, there have already been a few cases in which transit-timing variations betrayed the existence of an otherwise unseen planet, and the planet's existence was confirmed with the Doppler method.

In addition to providing a way to discover new planets, the transit-timing technique lets us test a basic prediction of planet formation theory: the planets' orbital planes should be nearly aligned with each other, as they are in the Solar System. If the orbits of two neighboring planets are aligned, the planets approach each other closely every time the inner planet laps the outer planet. Such close approaches lead to especially strong interplanetary gravitational forces. On the other hand, if the two orbits are inclined with respect to one another by a large angle, then a close encounter between the planets requires a coincidence: the inner planet needs to lap the outer planet when they are both near the line where the orbital planes intersect. Due to this effect and other longer-term gravitational effects, the observed pattern of transit-timing variations can sometimes be used to determine the angle between the planets' orbital planes. That's how Nesvorny and his colleagues were able to predict that the orbital plane of Kepler-46c is only slightly misaligned with that of Kepler-46b—just enough to prevent Kepler-46c from transiting in its current configuration.

Among the planetary systems for which this technique is applicable, we almost always find that the orbital planes line up to within 10°, and sometimes within 1°. This conclusion is also supported by statistical analyses of the frequency with which multiple planets around the same star have been detected using different methods. The Doppler and transit-timing methods can detect multiple planets even if their orbits are misaligned. But because the transit method requires planetary orbits to be viewed sideways, detecting multiple transiting planets around the same star is much more likely when their orbits are aligned. So, if there were a large population of misaligned systems, multiple-planet systems would be underrepresented in transit surveys compared to Doppler and transit-timing surveys—and, judging from current data, this is not the case. In this respect, the compact planetary systems described in chapter 5 resemble the Solar System, fulfilling a prediction of planet formation theory.

I admit to being a little disappointed. Misaligned orbits would be wonderfully weird. We do expect misaligned orbits to occur at least occasionally, as a result of close encounters and other gravitational interactions between planets—the same effects that have been blamed for the highly eccentric orbits and high obliquities seen in some systems. There is tentative evidence for large misalignments in a few systems, the most intriguing of which is HD 3167, where two planets appear to have perpendicular orbits, based on work by a team led by Vincent Bourrier of Geneva Observatory. Someday, I hope, we will confirm the existence of planetary systems with perpendicular orbits, and, even better, planets that revolve in opposite directions.

Puffball Planets

Another application of the transit-timing technique is to measure the masses of planets that are too lightweight to be measured with Doppler spectroscopy. This is possible because transit-timing observations are based on *photometry*, monitoring a star's overall brightness, which does not require as large a telescope as *spectroscopy*, monitoring the star's brightness at each of many wavelengths. Thanks to this advantage, transit-timing observations have revealed a new category of planets with abnormally low densities: the *puffball planets*.

For example, the Kepler-51 system has three transiting planets, two of which have radii of 7 $R_\oplus$ and 9 $R_\oplus$, making them comparable in radius to Saturn (9 $R_\oplus$)—but their masses are 2 $M_\oplus$ and 4 $M_\oplus$, making them much lighter than Saturn (95 $M_\oplus$). These peculiarities were discovered in 2014 in a transit-timing study led by Kento Masuda, then at the University of Tokyo. Both planets have average densities of about 0.03 g/cm^3, less than one-tenth of the density of Jupiter or Saturn and comparable to the density of Styrofoam. About a dozen other puffball planets have been found around other stars.

We don't yet know how to make sense of the puffball planets. The problem sounds just like the radius inflation problem for hot Jupiters, described in chapter 4, although puffball planets are much lower in mass and have wider orbits than hot Jupiters. One hypothesis holds that a planet becomes a puffball when its atmosphere is strongly heated by high-energy radiation emanating from a young star. This hypothesis is related to the theoretical scenario described in chapter 5, in which a mini-Neptune becomes a super-Earth when its gaseous atmosphere is driven off by high-energy radiation. Supporting this theory, the star in the Kepler-51 system is only 500 million years old, a youngster in comparison with the five-billion-year-old Sun. Perhaps by the time Kepler-51 reaches the Sun's age, the planets will lose their gaseous outer layers and settle down to become super-Earths. Another hypothesis is that at least some of the puffball planets are normal planets with rings. If their rings are thicker than Saturn's rings, they could block enough starlight during a transit to fool us into thinking that the planet itself is big and puffy. Very precise observations of transits, perhaps with the Webb Space Telescope, could allow us to distinguish between a puffy planet and a ringed planet.

CHAPTER SIX

Planetary Resonances

There's another interesting facet of the Kepler-51 system. The three known planets have orbital periods of 130, 85, and 45 days. If we divide each of those numbers by 45, we obtain 2.9, 1.9, and 1. In other words, the ratios of the three periods are nearly 3 to 2 to 1. Similar patterns are observed in some other multi-planet systems. For example, three of the planets in the GJ 876 system have orbital periods with ratios close to 4 to 2 to 1, and two planets in the K2–19 system have orbital periods very nearly in the ratio 3 to 2. The four planets in the Kepler-223 system show an even more intriguing pattern. Starting with the period of the first planet (7.38 days), you can obtain the periods of the other three planets by multiplying successively by $^4/_3$, $^3/_2$, and $^4/_3$.

You might be tempted to dismiss these patterns as meaningless coincidences. After all, the ratio of any two randomly chosen numbers might be close to a whole number or a simple fraction merely by chance, and besides, the orbital periods of the eight planets in the Solar System do not show any such tendency.[3] The exoplanet patterns can't be dismissed, though. Among the known exoplanetary systems, we observe mathematically simple ratios of orbital periods about twice as often as chance alone would dictate, and when they do occur, they have important ramifications for the fate of a planetary system and for the astronomers who are trying to study it.

3.   For the Solar System, there's a long history of trying to interpret the Titius–Bode law, which says the semimajor axis of the nth planet (in AU) is $0.4 + (0.3)^{n-2}$. This turned out to be naïve numerology: the formula doesn't work for Neptune and requires that we cheat for Mercury by inserting $n = -\infty$ instead of 1.

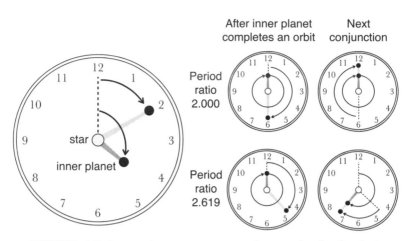

FIGURE 6.2. A two-planet system portrayed as two hands of a clock. Both hands start in the 12:00 position. If the inner hand moves twice as fast as the outer hand (top row), conjunctions always occur at 12:00. If the period ratio is more arbitrary (bottom row), conjunctions occur in many locations on the clock face (here, at 7:25).

To understand why, imagine looking down on a flat planetary system consisting of two planets on circular orbits. Draw a straight line from each planet to the star, thereby forming two hands of a clock (figure 6.2). The hands have different lengths and circulate at different rates. Obeying Kepler's third law, the inner hand circulates faster than the outer hand (unlike a real clock, in which the hour hand moves more slowly than the minute hand). Every time the inner planet laps the outer planet, the inner and outer hands line up, an event called a *conjunction*. During conjunctions, the planets are close together and the gravitational forces between them are especially strong, leading to small changes in the speeds and shapes of their orbits. The changes are small because the interplanetary forces are weaker than the star's gravitational pull.

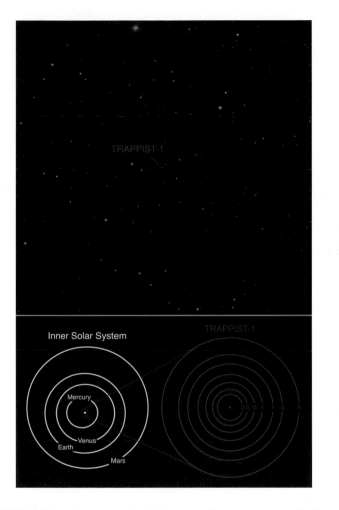

PLATE 1. (Top) An astronomical image spanning one-quarter of a degree on a side. The reddest star, labeled TRAPPIST-1, has seven known planets. The planets are too small and too close to the star to be seen in any image, but are known to exist because they periodically eclipse the star. (Bottom) Diagram of the orbits of the TRAPPIST-1 planets (red), compared to the much wider orbits of the innermost planets in the Solar System (yellow). Credit: Sloan Digital Sky Survey.

PLATE 2. Relative sizes of the planets. Jupiter is approximately ten times larger than Earth. Uranus and Neptune are approximately four times the size of Earth. The Sun (not shown) is approximately ten times larger than Jupiter. Credit: NASA/JPL.

PLATE 3. Artists' depictions of disks surrounding young stars. Not shown is the layer of gaseous hydrogen and helium, which is thin and dense near the star, and becomes less dense and vertically thicker farther away. Planets are thought to form within such disks, starting from grains of dust. Credit: NASA/JPL (top); ESO/L. Calçada (bottom).

Spectrum of starlight, with spectral absorption lines

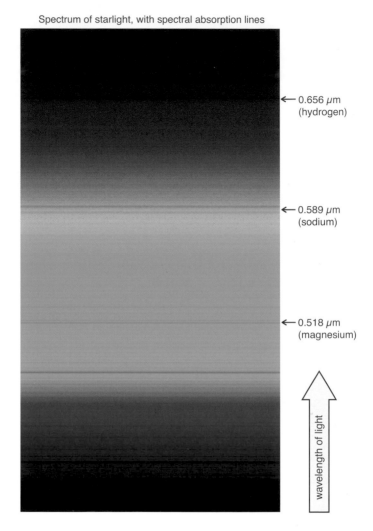

←— 0.656 μm
(hydrogen)

←— 0.589 μm
(sodium)

←— 0.518 μm
(magnesium)

wavelength of light

PLATE 4. A spectrograph disperses starlight into a rainbow of colors, using a prism or a finely ruled reflective surface called a diffraction grating. The rainbow contains dark lines corresponding to the wavelengths of light that are absorbed by substances in the star's atmosphere. For example, sodium absorbs light with a wavelength of 0.589 μm (0.589 millionths of a meter). Credit: NASA/STScI.

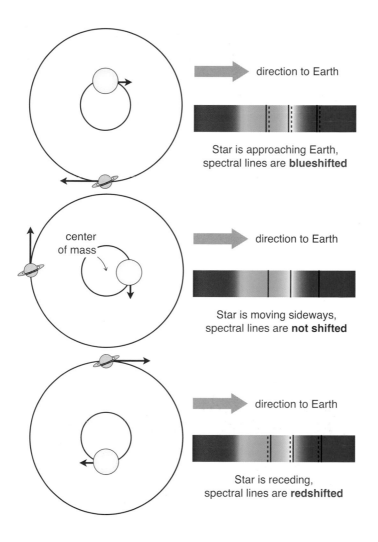

direction to Earth

Star is approaching Earth,
spectral lines are **blueshifted**

center
of mass

direction to Earth

Star is moving sideways,
spectral lines are **not shifted**

direction to Earth

Star is receding,
spectral lines are **redshifted**

PLATE 5. Conceptual illustration of the Doppler method for detecting exoplanets. The star's motion around the center of mass causes Doppler shifts in the wavelengths of the star's spectral absorption lines.

PLATE 6. An artist's conception of the WASP-12 planetary system. Gas is streaming away from a bloated, egg-shaped hot Jupiter. The gas swirls around the star in an accretion disk, and ultimately falls onto the star. Credit: NASA, ESA, and G. Bacon (STScI).

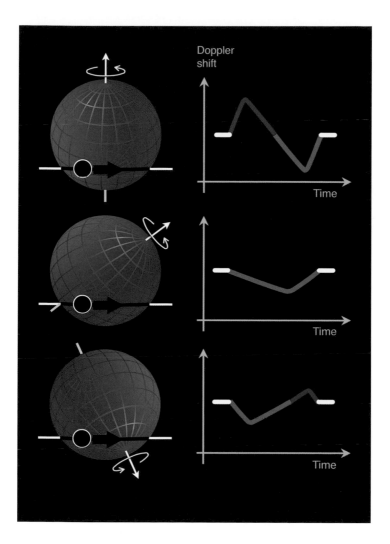

PLATE 7. The Rossiter–McLaughlin effect. At left are three stars with rotation axes pointing in different directions. Because of rotation, half of each star is blueshifted and half is redshifted. Usually these effects cancel out, but during a planetary transit, there is a slight imbalance that can be detected as a net Doppler shift. The corresponding Doppler shifts are shown at right.

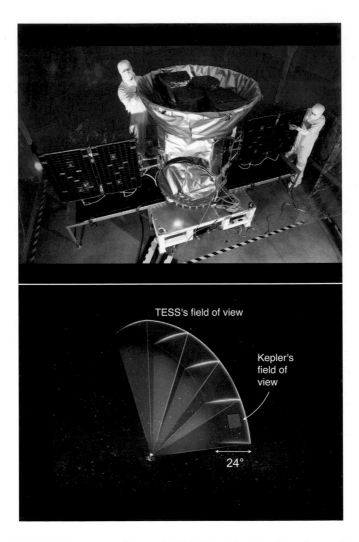

TESS's field of view

Kepler's field of view

24°

PLATE 8. The top panel shows NASA's Transiting Exoplanet Survey Satellite (TESS) before it was launched in 2018. TESS uses four small optical telescopes to monitor nearby stars for transiting planets. The four telescopes point in different directions, as illustrated in the bottom panel, forming a field of view larger than that of NASA's earlier Kepler space telescope. Credit: Northrup Grumman (top); NASA/GSFC (bottom).

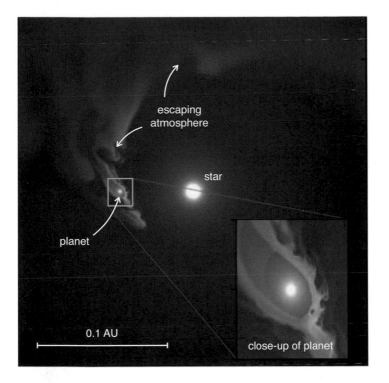

PLATE 9. Computer simulation of the WASP-107 system. Intense radiation from the star causes some of the planet's atmosphere to escape into the surrounding space and trail behind the planet. The color indicates the density of gas. Credit: M. MacLeod and A. Oklopčić.

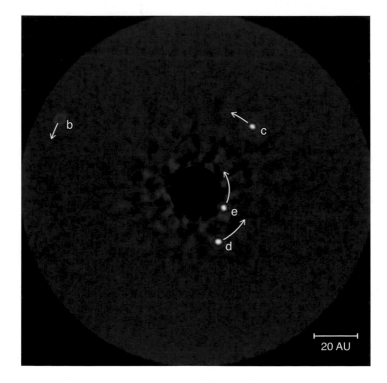

PLATE 10. The HR 8799 planetary system. Light from the central star has been suppressed by a coronagraph, revealing four faint planets. The arrows indicate the observed motion of each planet over seven years. The orbit of planet "c" has approximately the same size as Neptune's orbit around the Sun. Credit: NRC–HIA, C. Marois, and Keck Observatory.

PLATE 11. The Beta Pictoris planetary system. The top panel is an image in which light from the central star has been suppressed, revealing a dusty disk seen edge-on, as well as two faint planets (labeled "b" and "c"). The bottom panel is an artist's conception of the view from a different angle. Credit: GRAVITY Collaboration; Axel M. Quetz, MPIA Graphics Department.

Orbits of Jupiter, Saturn, Uranus,
and Neptune on the same scale

PLATE 12. Image of the protoplanetary disk surrounding the young star HL Tauri, based on data from the Atacama Large Millimeter-wave Array. This is a "false-color" image in which color conveys the intensity of millimeter-wave radiation. For comparison, the Solar System is depicted on the same scale and from the same viewing angle. Credit: ALMA (ESO/NAOJ/NRAO).

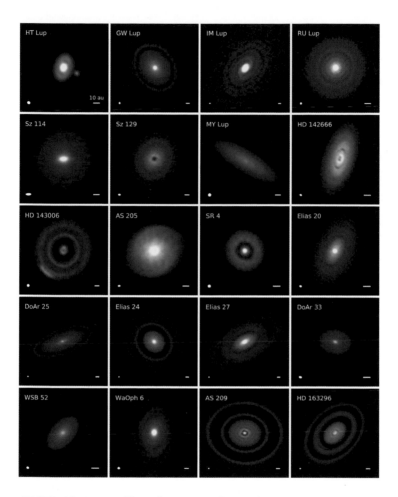

PLATE 13. ALMA false-color images of protoplanetary disks around 20 young stars. Some are disks with dark lanes, others have rings, and a few show spiral patterns reminiscent of galaxies. In each panel, the scale bar in the lower right corner spans 10 AU, and the ellipse in the lower left corner indicates the size and shape of the smallest resolvable details in the image. Credit: S. Andrews and the DSHARP team.

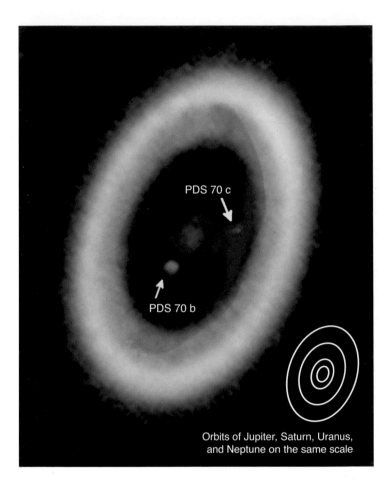

PLATE 14. False-color image of the ring of protoplanetary material surrounding the young star PDS 70. The white-orange color conveys the intensity of millimeter waves coming from the dust in the ring. Optical and infrared images (red and blue) show two newborn planets. Credit: ALMA (ESO/NAOJ/NRAO) A. Isella; ESO.

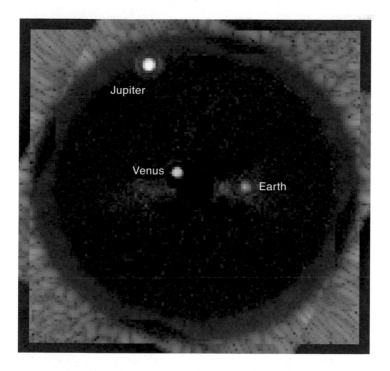

PLATE 15. Simulated image of the inner Solar System as it would appear from 40 light-years away, using a hypothetical 15-meter space telescope equipped with a coronagraph. Light from the Sun has been suppressed by a factor of 10 billion. Credit: R. Juanola Parramon, N. Zimmerman, and A. Roberge (NASA/GSFC).

PLATE 16. Artist's conception of a starshade, a possible solution to the problem of making images with a high enough contrast to detect an Earth-like planet. The shade is engineered to block light from the star while allowing light from nearby planets to reach a space telescope. In reality, the telescope would be thousands of kilometers behind the starshade. Credit: NASA/JPL–Caltech.

Let's imagine watching the planets go around the star, starting from a conjunction. We'll orient the clock face so that the conjunction occurs at the "12:00" position, when both hands are pointing straight up. We need to pay attention to where the next conjunctions occur.

If the ratio of the orbital periods is not a whole number or simple fraction, then conjunctions occur at many different locations on the clock face. For example, if the period ratio is 2.619 (chosen arbitrarily), the next five conjunctions occur at the locations 7:25, 2:49, 10:14, 5:39, and 1:04. Because the conjunctions occur at so many different locations along each planet's orbit, the small disturbances in their orbits tend to cancel each other out in the long run.

Things are different if the ratio of orbital periods is a whole number or simple fraction. For example, suppose the inner hand moves exactly twice as fast as the outer hand. In the time it takes the inner hand to go all the way around and come back to 12:00, the outer hand gets only as far as 6:00. By the time the slowpoke outer hand goes from 6:00 back to 12:00, the inner hand has gone all the way around a second time. The two hands meet again at the 12:00 position, forming a conjunction. Since the hands are now back where they started, the whole process repeats. So, in this case, the only place that conjunctions occur is the 12:00 position. Likewise, if the period ratio is 3 to 2, then in the time it takes the inner hand to complete three cycles, the outer hand completes exactly two, and both hands meet at the 12:00 position.

In these cases, the gravitational disturbances caused by strong interplanetary forces always occur at the *same location* within each planet's orbit. This allows minor disturbances to accumulate into major disturbances. After hundreds, thousands, or billions of years, the planets' orbits can be dramatically reshaped—much more so than when the period ratio is not a whole number or simple fraction.[4]

Special period ratios are examples of a widespread physical phenomenon called *resonance*. Resonances can exert a stabilizing effect on a planetary system, or a destabilizing effect, depending on the details. For example, in the Solar System's asteroid belt, which extends between Mars and Jupiter, there are zones nearly devoid of asteroids and other zones with especially large congregations of asteroids. These zones correspond to orbital periods that are in different resonances with Jupiter's orbital period. Another example is furnished by Pluto and Neptune. They have elliptical orbits that nearly intersect each other, which sounds like a recipe for disaster. However, they are in a 3 to 2 resonance, such that conjunctions only occur when Neptune is at its closest approach to the Sun and Pluto is farthest away. This keeps the two planets at a safe distance from one another.

This protective resonant arrangement seems too good to be true. It's as though the orbital periods of Neptune and Pluto, and similar systems, were selected as a survival strategy. As I noted earlier, it's not just a coincidence. Resonances and near-resonances (near enough to have detectable effects on the planetary orbits) occur in about 10% of the known exoplanetary systems, which is about twice as often as one would expect if period ratios were drawn randomly from a continuous distribution. With all due respect to Isaac Newton, we don't think protective resonances are examples of divine interven-

---

4. Conjunctions always occur at 12:00 for any period ratio of the form $n+1$ to n, where n is a whole number. Ratios of the form $n+2$ to n (and not $n+1$ to n), such as 5 to 3, lead to milder effects because conjunctions occur at both 12:00 and 6:00, allowing for partial cancellation of the gravitational disturbances. For fun (if this is your idea of fun), try thinking about the general case $n+m$ to n.

CHAPTER SIX

tion. But how do these special mathematical relationships arise from inanimate objects and the laws of physics?

For the exoplanetary systems, at least, the best available answer is disk-driven migration. As discussed in chapter 4, disk-driven migration is one of the ideas that was proposed to explain the existence of short-period giant planets. In this scenario, a giant planet forms at a distance of a few AU, beyond the snow line of the protoplanetary disk. Afterward, gravitational interactions with the protoplanetary disk cause the planet to spiral inward and arrive in a much smaller orbit. A smaller dose of disk-driven migration might explain the resonant exoplanetary systems. Suppose two planets initially have periods in a ratio of, say, 2.619. Each of them interacts with the protoplanetary disk and begins migrating inward. If the outer planet migrates faster—because it is less massive and easier to move around, or for some other reason—its orbital period shrinks faster than the orbital period of the inner planet, and the period ratio decreases. As the period ratio approaches 2, the planetary conjunctions start occurring in nearly the same location every time, allowing the effects of interplanetary forces to build up. Sometimes the result is catastrophic. The planets throw each other onto highly eccentric and inclined orbits which eventually intersect, leading to collisions or ejections. Sometimes, though, if the migration process is slow enough, the planets become locked into a resonance that prevents catastrophic interactions, as is the case for Pluto and Neptune. In this sense, the existence of exoplanetary resonances is evidence for at least occasional disk-driven migration.

There are also practical implications of resonances for astronomers. The planets in resonant and nearly resonant systems show unusually large transit-timing variations, because of the repetitive character of the interplanetary forces and the resulting growth of minor perturbations into major ones. This allows us to measure the timing variations with high precision and extract unusually detailed information. Nature has been kind to us, again, by arranging for the most prominent possible displays of the gravitational forces between planets.

My favorite example is the extraordinary exoplanetary system named TRAPPIST-1. As noted in the introduction, TRAPPIST-1 is a diminutive red star with seven transiting planets (plate 1). The planets are all comparable in size to the Earth, ranging in radius from 0.8 $R_\oplus$ to 1.1 $R_\oplus$, and *every one of them* is in resonance. The resonances are even more intricate than the examples given earlier because they involve three planets at a time instead of two.[5] Because of this interlocking chain of resonances, a team led by Eric Agol was able to use transit-timing variations to measure the planets' masses. Combining this information with the measurements of the planets' radii, the team calculated the planets' overall densities, which range from 4.1 g/cm^3 to 5.4 g/cm^3. Their densities are all a little lower than Earth's overall density of 5.5 g/cm^3, suggesting that the TRAPPIST-1 planets have a somewhat different composition than Earth—probably a higher abundance of water and other lightweight substances.

The transit data also showed that the planets' orbits are almost exactly circular, with eccentricities smaller than 0.01. The orbits are aligned with each other to within 0.2°, making the system flatter than the Solar System (and flatter than a pancake). I find it astonishing and inspiring that such a wealth of

---

5. For every triplet of adjacent planets, the three periods obey an equation $n/P_1 - (n+m)/P_2 + m/P_3 \approx 0$, where n and m are small integers. This type of *Laplace resonance* prevents simultaneous conjunctions of all three planets. Jupiter's moons Io, Europa, and Ganymede are in a Laplace resonance.

CHAPTER SIX

information about a planetary system 40 light-years away can be obtained by decoding an erratic pattern of fading and brightening of a faint red dot in the sky. In his poem "Auguries of Innocence," William Blake invited us To See a World in a Grain of Sand. The TRAPPIST-1 system invites us To See Seven Worlds in a Glimmer of Light.

We can also *hear* the resonances of TRAPPIST-1 and other systems, by relying on an analogy between resonance in planetary systems and musical harmony. When a piano string vibrates, the period of vibration determines the wavelength of the resulting sound waves and the pitch of the musical note. Shorter periods (faster vibrations) correspond to higher pitches. When the main vibration periods of two strings are in the ratio 2 to 1, they produce notes that harmonize perfectly. Even though the notes have different pitches, to our ears they blend together so well that they are assigned the same letter in musical notation. They're said to be an *octave* apart, such as middle C and the next higher C. Strings that have vibration periods in the ratio 3 to 2 also sound pleasant when played together. The resulting harmony is called a *perfect fifth*, such as between C and G within the same octave. A period ratio of 4 to 3, as exists between C and F within the same octave, is a *perfect fourth*. These simple period ratios play fundamental roles in the Western musical tradition.

Just for fun, Matt Russo, Daniel Tamayo, and Andrew Santaguida created an audio version of TRAPPIST-1 by assigning a piano key to each planet, such that the vibration periods of the piano strings have the same ratios as the planets' orbital periods. The outermost planet was designated middle C, causing the other planets to chime in at higher pitches ranging up to the second-highest D on the piano keyboard. In their

YouTube video,[6] the planets of TRAPPIST-1 circulate like a seven-handed clock. Each planet's note is sounded when it completes a full orbit. The resulting composition is not Mozart, but it is somewhat musical.[7] Trying to do the same for a planetary system far from resonances would cause a cacophony and make you rush to press the mute button.

One of the most intriguing resonant exoplanetary systems is Kepler-36. Of all the exoplanetary systems that have been thoroughly investigated, Kepler-36 has the most closely spaced pair of planets. The planets' orbital periods are nearly in the ratio 7 to 6, and their orbital distances differ by only 10%. During conjunctions, the two planets approach each other as closely as 0.013 AU. Stargazers on the inner planet, viewing the outer planet in their sky over the course of a few months, would see a point of light grow into a brilliant globe nearly three times larger than our full Moon.

In music, playing two notes with vibration periods in a ratio of 7 to 6 sounds terrible. The equivalent on a piano keyboard would be a C and a slightly sharp D, a dissonant interval. However, in gravitational dynamics, a dissonant period ratio still counts as a mathematical resonance. Because the two planets of Kepler-36 approach each other so closely, they pull on each other with unusual vigor, causing large and erratic transit-timing variations. The pattern of variations depends so sensitively on the locations of the planets that it's difficult to make accurate long-term predictions of future transit times.

---

6. Try youtu.be/7i8Urhbd6eI, or search for "The Song of a Solar System: TRAPPIST-1."

7. There's a long tradition of associating musical notes with planetary orbits. Followers of the ancient Greek philosopher Pythagoras wrote about the "music of the spheres," and Kepler wrote a book called *Harmonices Mundi* (The Harmony of the World).

CHAPTER SIX

Unpredictability is the hallmark of *chaos*, in the technical mathematical sense of the word. A physical system is chaotic whenever two copies of the same system, differing only slightly in their initial conditions, would have strongly divergent futures. This is also known as the "butterfly effect." The weather is the classic example of a chaotic system: metaphorically speaking, the flapping of a butterfly's wings in Brazil could eventually cause tornadoes in Texas. In the same metaphorical spirit, we might say that the gravitational effect of a butterfly flitting past one of the Kepler-36 planets would eventually cause the planets to be in completely different locations along their orbits than if the butterfly had flown elsewhere.

Kepler-36 was not the first planetary system known to exhibit mathematical chaos. Most planetary systems, including the Solar System, are technically chaotic. That's why Newton wondered if supernatural forces were needed to preserve the stability of the Solar System, and why eighteenth-century physicists experienced such headaches when they tried solving the problem of three or more gravitationally interacting bodies. Starting in the late 1980s, theorists conducted computer simulations to study the future of the Solar System and found that chaos prevents accurate predictions beyond about 100 million years. Over that time, an error of 15 meters in our knowledge of Earth's initial position would lead to an error in the predicted position of 150 million kilometers. We can't even be completely certain that all the planets will survive the next few billion years. In about 1% of the simulations conducted in 2009 by Jacques Laskar, an expert in gravitational dynamics at Paris Observatory, Mercury suffered a terrible fate by either falling into the Sun or colliding with Venus—and in one terrifying case, Mercury smashed into the Earth.

For the Solar System, such cataclysmic events are improbable, but for Kepler-36, the threat is imminent. The system is on the brink of instability. If the planets were just a little more massive, or if their orbits were a little more eccentric, the system would rapidly undergo drastic rearrangements or planetary collisions. Chaos and unpredictability erupt on a timescale of a few decades, a million times faster than in the Solar System. Kepler-36 shows the most rapid descent into unpredictability of any known planetary system. We do not know what caused this planetary system to exist in such a fragile state.

Newton used the regular motions of the planets of the Solar System to help him to infer the universal laws of motion and gravity, and thereby propel the scientific revolution. What would it be like for an exoplanetary analog of Newton to try to grasp the laws of physics while living on a planet within a strongly chaotic system? Would an exo-Newton arrive at the theory of mathematical chaos centuries before the equivalent point in our civilization—or would the scientific revolution be stalled by the lack of a straightforward celestial physics demonstration? A science fiction novel, *The Three-Body Problem* by Cixin Liu, explored the ramifications of chaotic planetary motion on an extraterrestrial civilization. This novel was published in 2008, four years before the paper reporting the discovery of Kepler-36. It's a good example of science catching up with science fiction.

Lucas et al. (1977)

My favorite example of a real exoplanet with a sci-fi predecessor is Kepler-16b. It's a Saturn-sized transiting planet with an orbital distance of 0.7 AU. By themselves, these are not remarkable properties. The remarkable aspect of Kepler-16b is that it orbits *two stars*. The paper reporting the discovery of this system (in which I played a minor role) was published in 2011, but George Lucas got there first. He reported a system with a similar geometry in 1977 in a publication called *Star Wars*. Think of Luke Skywalker, gazing longingly across the desert landscape of Tatooine, watching two suns sink toward the horizon.

As I've mentioned, binary stars are common throughout the galaxy. The separation between the two stars varies widely from system to system, ranging from 0.01 AU to 10,000 AU. It is not well understood why binary stars are so common. Some of them probably form when a gravitationally collapsing region of a gas cloud splits into two blobs at a late stage of the collapse, each of which then collapses further to form a star. Others may form when the gaseous disk surrounding a young star is so massive that it becomes gravitationally unstable. The swirling disk develops an internal vortex, a smaller and denser whirlpool of gas that develops into a second star.

Even before discovering Kepler-16, we knew that some binary star systems have planets, but in the previously known cases, the planet orbits closely around one of the two stars and the second star travels in a wide orbit around the first star and its planet. In the Gamma Cephei system, for example, the two stars have an elliptical orbit with a semimajor axis of 20 AU, and the larger star has a planet on a nearly circular orbit with a radius of 2 AU. This is the planet I described in chapter 3 that was discovered (or maybe just detected?) by Gordon Walker and his group in the 1980s.

In contrast, Kepler-16b is a *circumbinary* planet, defined as a planet whose orbit surrounds a pair of stars, rather than just one star. Kepler-16 is the most well-studied of the 15 known

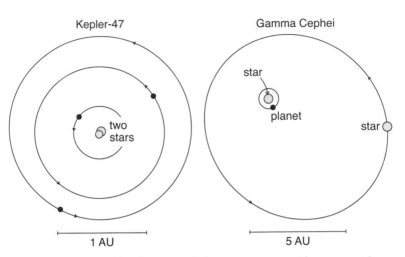

FIGURE 6.3. Orbit diagrams of planetary systems with two stars. In Kepler-47, a circumbinary planetary system, the two stars orbit each other, and three planets orbit around both stars. In Gamma Cephei, the second star is orbiting around the first star and its planet.

examples of circumbinary planets. The masses of the two stars are 0.2 $M_\odot$ and 0.7 $M_\odot$. The orbital planes of all three bodies—star, star, and planet—are aligned to within 0.5°. From our vantage point in the galaxy, we are viewing the system from the side. We see the stars periodically eclipsing one another, and we also see the planet transiting across both stars. The transits and the eclipses were detected with the Kepler space telescope. The timing and other characteristics of the eclipses are reproduced well by a simulation of gravitational dynamics in a three-body system, leaving no room to doubt that this is a genuine case of a Tatooine-like planet. Another well-studied system, Kepler-47, features *three* giant planets, all of which circle around a binary star (figure 6.3).

Young binary stars with separations smaller than a few AU are sometimes observed to be at the center of a single circular disk of gas and dust. These *circumbinary disks* are similar to the protoplanetary disks that surround single young stars, but they have large central openings that accommodate both stars. Before the discovery of Kepler-16, it was an open question whether planets could form within a circumbinary disk. Planet formation is supposed to be a bottom-up process, with dust grains sticking together to make rocks that glom together to make boulders and eventually planets. Within a circumbinary disk, though, the material would be constantly stirred up by the fluctuating gravitational forces from the two stars moving around at the center of the disk. Could planets grow when the raw materials are being perpetually jostled? The search for transiting circumbinary planets, led by Laurance Doyle of the SETI Institute in Mountain View, California, was proposed as a "stress test" of the planet formation process. Truth be told, we were at least as excited about confirming science fiction as we were about testing the theory of planet formation.

Spectroscopy 101

Just as gravitational dynamics is a tool for studying the orbital characteristics of exoplanets, spectroscopy is a tool for studying their atmospheres. To understand a planet, measuring the planet's mass and radius isn't enough: we must know something of its atmosphere. In terms of mass and radius, Venus and Earth are basically twins. Alien astronomers using only the Doppler and transit methods to study the Solar System would consider Venus and Earth to be very similar planets. We know better. Venus's surface is hotter than a soldering iron because of the smothering greenhouse effect of its thick carbon dioxide atmosphere, which exerts a hundred times more pressure than Earth's atmosphere. So, if we detect an exoplanet with the same size and mass as the Earth, how could we tell whether it is a big blue marble or a fire-and-brimstone hellhole?

We can take a hint from Venus itself. Venus's atmosphere was first detected in 1761, during one of its rare transits across the Sun. Mikhail Lomonosov, a Russian polymath, reported that when Venus first moved across the rim of the Sun, a thin luminous ring developed around the black circle of Venus's shadow. This luminous ring was sunlight being refracted through Venus's atmosphere. We can also use transits to detect the atmospheres of exoplanets, although not in the same way. Instead of looking for the effects of refraction, we perform spectroscopy, a crucial tool used throughout astrophysics to determine the physical conditions within heavenly bodies.

The essential predicament of astronomy is that we see fascinating objects through our telescopes, but we cannot bring them into a laboratory for further assessment. For a long time, this limitation made it seem impossible that humankind could ever know the true nature of astronomical bodies. Speculating about the composition of stars and planets must have seemed like speculating about how many angels can dance on the head of a pin. Starting in the mid-nineteenth century, innovations in spectroscopy allowed these speculations to be converted into scientific investigations. As it turns out, we can learn a lot about the composition of an object based on the spectrum of the light that it emits or absorbs. A spectrograph, attached to a telescope, makes the universe a little more like a laboratory.

CHAPTER SIX

Recall that certain colors appear to be missing from the Sun's spectrum. For example, there's a conspicuous dark line in the red part of the Sun's spectrum at a wavelength of 0.656 μm (0.656 millionths of a meter). Spectral absorption lines act as tick marks that allow us to measure Doppler shifts precisely, but unlike the ticks on a ruler, the dark lines are not regularly spaced. They don't look random, either. In some parts of the spectrum, there are clusters of dark lines, while in other parts the continuous rainbow is hardly interrupted at all. It's as though the dark lines are a message written in a secret code.

In fact, the dark lines *are* a message, sent to us by the atoms in the Sun's outer layers. The atoms are broadcasting information about their identities and the local temperature and pressure. Hundreds of the Sun's spectral absorption lines were documented in 1814 by Joseph von Fraunhofer, but the message within the lines wasn't decoded for more than a century afterward. That's because the message is written in the language of quantum mechanics, which wasn't understood until the 1920s.

Before quantum mechanics, atoms were pictured as nanoscale models of the Solar System, with the nucleus representing the Sun and the electrons standing in for planets. Indeed, there are some genuine similarities. In both atoms and planetary systems, the strength of the force that holds the system together—electricity for atoms and gravity for planets—varies inversely as the square of the distance between the objects, and in both cases, the orbiting objects have much smaller masses than the central object. There are also profound differences between planets and electrons, because electrons are fundamental particles, and because the electrical force is intrinsically much stronger than the gravitational force.[8] Together, these properties cause atoms to be much smaller than planetary systems and to display the mind-bending aspects of quantum mechanics in their full glory.

A few quantum-mechanical facts are crucial for understanding spectroscopy. First, while planets don't abruptly change the sizes of their orbits, electrons do. An electron inside an atom can absorb the energy from light or other types of electromagnetic radiation, causing a sudden change in the size and shape of its orbit. Usually, absorbing energy causes an electron's orbit to widen. The reverse happens, too. An energetic electron can emit a photon—a bundle of electromagnetic energy—causing the electron's orbit to contract. Second, while a planet's orbit around a star can be of any size, an electron's orbit is limited to a restricted list of possible sizes and energies. The implication is that an electron cannot absorb *any* photon that might be passing by, and it cannot emit a photon with *any* energy it might wish. An electron in an atom can only absorb or emit a photon if the photon's energy is just the right amount to bring the electron from its current orbit to a different orbit in the list of allowed orbits.[9] Importantly, the list of allowed orbits is specific to each type of atom. And although I've been referring to atoms, the same applies to ions (atoms with missing or extra electrons) and molecules

8. Much, much stronger. For example, the electrical attraction between an electron and a proton is 2×10^{39} times stronger than the gravitational attraction between them. Just for fun, let's write that out:
    2,000,000,000,000,000,000,000,000,000,000,000,000,000.
9. It can also absorb a photon with enough energy to remove the electron from the atom altogether. So, there is both a list of discrete energies and a threshold above which a continuous range of energies can be absorbed.

CHAPTER SIX

(groups of atoms stuck together). Hydrogen, helium, water, methane, and oxygen all have different lists of allowed orbits and energies.

With these facts in mind, we can understand the basics of spectroscopy. Suppose we send an intense beam of white light through a glass chamber containing a dilute gas. White light contains photons with a wide range of energies. Most of the photons have energies that cannot be readily absorbed by atoms in the gas. Such photons pass straight through the gas and exit the chamber, where they are detected by our spectrograph. On the other hand, if a photon happens to have an appropriate energy, it can get absorbed as it passes through the gas. The emerging beam of light will be missing some of the photons with that special value of energy. Because a photon's energy determines its wavelength, our spectrograph will show that the emerging beam of light is darker at the corresponding wavelength.[10]

That's how to observe spectral absorption lines in a physics laboratory. To observe them in the heavens, we use a telescope to send starlight through a spectrograph. The intense radiation propagating upward from deep within the star's hot interior plays the same role as the intense light beam in the laboratory experiment. The light passes through the dilute outermost layers of the star's atmosphere, which act like the container of dilute gas in the laboratory. The stream of photons that emerges

10. A subtlety: when an atom absorbs a photon, it often emits an identical photon shortly afterward. But absorption and emission don't cancel each other out, because the emitted photons fly in many directions and few of them rejoin the original beam of light. If we observe the gas from the side, away from the direction of the incident beam, we see *only* the emitted photons, which form a spectrum of bright lines (an *emission spectrum*) instead of dark absorption lines.

166

from the star and eventually lands in our spectrograph is missing the photons that were filtered out by the constituents of the star's atmosphere.

Spectroscopy of Hot Jupiters

By matching the pattern of dark lines in a spectrum with the lists of energies that can be absorbed by various substances, we can identify which substances are present in the atmosphere of a star. A more detailed analysis can also reveal the temperature and pressure of the star's atmosphere. For exoplanets, though, there is a problem. As we've seen, it's almost impossible to separate the light from a star and its planets in an image—their light blends together in our telescopes. With the star's bright glare overpowering the planet's light, how can we detect the planet's spectrum?

The trick is to monitor the star's spectrum before, during, and after a planetary transit. If we could view the transit up close, we would see a small black circle—the planet's silhouette—moving across the larger and brighter circle formed by the star. Looking even more closely, we would see that the black circle has a fuzzy edge, where a small fraction of the starlight is filtering through the planet's partially transparent outer atmosphere. Suppose the atmosphere contains sodium atoms, which preferentially absorb light with a wavelength of $0.589\ \mu m$. If we could view the transit through glasses that only transmit light with a wavelength of $0.589\ \mu m$, the absorption by sodium would cause the planet's outer atmosphere to look darker. The planet's dark silhouette would have a larger radius, and it would block more light than it does at neighboring wavelengths (figure 6.4).

CHAPTER SIX

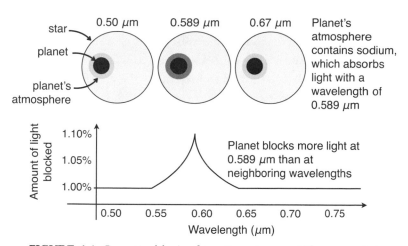

FIGURE 6.4. Conceptual basis of transit spectroscopy. When a transit is observed at the wavelength of an absorption line in the planet's atmosphere, the planet blocks more light than usual. By observing at many wavelengths, the spectral lines can be identified, providing information about the atmosphere's composition.

From afar, we cannot see the planet's silhouette and its fuzzy edge. Nevertheless, by recording the star's spectrum during a transit, we can identify the wavelengths at which the star appears slightly fainter because the planet's outer atmosphere is absorbing more light. We can tell the difference between these planetary absorption lines and the star's own absorption lines by observing the star before and after the transit, when we can see the star's unaltered spectrum.

The first detection of an exoplanet's atmosphere was reported in 2001, just a couple of years after the first discovery of a transiting planet, HD 209458b. The effort was led by David Charbonneau at Harvard University, whose team used a spectrograph aboard the Hubble Space Telescope. His Harvard colleagues Sara Seager and Dimitar Sasselov had calculated that sodium atoms would produce the darkest absorption line, based on the expected conditions in the atmosphere of a hot Jupiter—and indeed, sodium was detected.[11] Since then, the same method has been used to investigate the atmospheres of about 50 other exoplanets, most of which are hot Jupiters. Sodium is sometimes detected, along with its chemical cousin potassium. These atoms are not especially abundant. Hydrogen is far more abundant; there might be only one sodium atom for every 100,000 hydrogen atoms. But in the conditions of hot Jupiter atmospheres, hydrogen doesn't have absorption lines at visible wavelengths,[12] while sodium happens to have a very dark line in the yellow-orange part of the spectrum. Likewise, once it became possible to obtain reliable measurements at infrared wavelengths, astronomers detected water vapor in the atmospheres of some hot Jupiters, which might only exist in small quantities but produces dark absorption features in the infrared part of the spectrum. I invite you to reflect on the fact that by analyzing the light of a distant star, we can not only reveal the existence of a giant planet, but also discern that the planet's atmosphere is a little steamy.

The focus on hot Jupiters, so far, is mainly for practical reasons. Close-orbiting giant planets have the most readily detectable atmospheres. Their short orbital periods allow for frequent opportunities to observe transits, and their large sizes lead to relatively large atmospheric signals—though I

11.   In 2022, a team led by Giuseppe Morello revisited the 2001 data and concluded that the claim to have detected sodium had not been completely justified—another example of the difficulty of establishing who was "first" to make a discovery (chapter 3).

12.   Hydrogen's red absorption line at 0.656 μm, mentioned earlier, is only prominent when the temperature is about 10,000°C, because it requires the electron to start in the second energy level as opposed to the lowest energy level. Hot Jupiters, typically between 700°C and 2,200°C, are too cold.

CHAPTER SIX

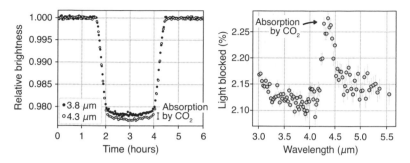

FIGURE 6.5. Evidence for carbon dioxide in the atmosphere of the hot Jupiter WASP-39b, based on observations with the Webb Space Telescope. The left panel shows the star's brightness versus time at two different wavelengths. The planet blocked slightly more light at 4.3 μm than at 3.8 μm, because of absorption by carbon dioxide at 4.3 μm. The right panel shows the full transit spectrum—the amount of light blocked at each of many wavelengths—in which the extra absorption at 4.3 μm is more obvious.

hasten to add that even for hot Jupiters, these are challenging measurements, with typical brightness variations smaller than 0.1% between different wavelengths in the spectrum (figure 6.5).

Along with these practical reasons, planetary scientists are drawn to hot Jupiters because of the extreme and exotic conditions of their atmospheres. Hot Jupiters are hot enough to cause many familiar substances to vaporize, leading to a wider variety of possible atmospheric gases than are present in the frigid atmospheres of Jupiter and Saturn. For example, some hot Jupiters show evidence for gaseous titanium oxide, whereas on Earth, titanium and oxygen combine to form a powder used in sunscreen and white paint. As another example, a star's ultraviolet radiation is expected to cause complex chemical reactions in a hot Jupiter's atmosphere, producing hazes and clouds unlike any that are found in the Solar System.

One of the most exotic atmospheres that has been probed is that of WASP-76b, a hot Jupiter nearly twice the size of Jupiter itself. As noted in chapter 4, there is no convincing theory to explain the swollen size of WASP-76b and the other inflated hot Jupiters. Regardless of the reason, the planet's distended atmosphere is convenient for transit spectroscopy, because it forms a larger region through which the starlight is filtered when the planet is transiting. This increased opportunity for absorption, along with the brightness of the star, make WASP-76b's atmosphere uniquely accessible. The inventory of elements that have been found in the atmosphere so far includes lithium, sodium, magnesium, calcium, vanadium, chromium, manganese, nickel, strontium, and iron.

The transit spectrum of WASP-76b has a sumptuous sodium absorption line that can be observed in fine detail. The range of wavelengths spanned by the sodium line is larger than would be expected if the planet's atmosphere were calm and undisturbed. According to a team of astronomers led by Julia Seidel and David Ehrenreich, at Geneva Observatory, the broad absorption line is evidence of a windy atmosphere. In chapter 4, I mentioned that one of the quirks of hot Jupiters is that they are unevenly heated, with one side getting roasted and the other side facing the cold emptiness of space. The uneven heating is expected to drive planet-wide wind patterns with outrageous speeds. An atmosphere in rapid motion would show a wide range of Doppler shifts, thereby broadening the range of wavelengths spanned by a spectral absorption line. The heat from the star appears to be powering a wind that circulates around WASP-76b with a speed of 5 kilometers per second, in rough agreement with theoretical expectations.

CHAPTER SIX

The absorption lines from iron atoms tell an even more remarkable story. Absorption by iron wasn't detected throughout the entire transit; it was only detected during the latter part of the transit. According to Ehrenreich and his colleagues, the reason is that iron vapor is only present in the hottest part of the planet's atmosphere. The side of the planet directly facing the star is hot enough to vaporize iron. The strong planetencircling wind carries the intense heat and the iron vapor around the planet to the nightside, where the atmosphere cools off, causing the iron vapor to condense into liquid droplets. When the wind carries the iron droplets back around to the dayside, the atmosphere warms up and by "noontime" (when the star is directly overhead) the iron has vaporized again.

Figure 6.6 illustrates this idea. At the start of the transit, when the planet's silhouette begins crossing in front of the rim of the star, the starlight filters through the relatively cool part of the atmosphere—the part that has just blown over from the nightside. Because the iron is in liquid form, rather than gaseous, the iron cannot be detected through spectroscopy. That explains why Ehrenreich's team didn't see iron, at first. At the end of the transit, the starlight filters through the part of the atmosphere that is downwind of the hottest part of the planet's atmosphere. There, the iron is vaporized, making it detectable with spectroscopy.

This is not the only possible interpretation of the data. Computer simulations of the planet's dynamic atmosphere have been undertaken by other groups who have come to different conclusions about the reason for the changes in the iron absorption line. Observations with the Webb Space Telescope are expected to clarify the matter. I hope that Ehrenreich's team turns out to be correct, because I love the idea that on WASP-76b, it's always raining iron.

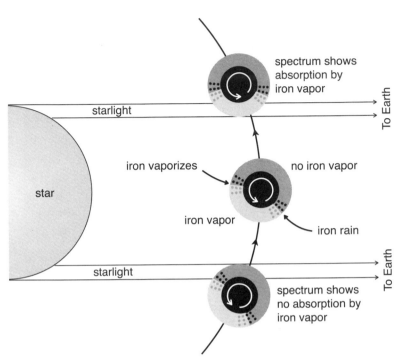

FIGURE 6.6. The iron-rain hypothesis for WASP-76b. The planet moves from bottom to top, and Earth is a long way off to the right. The planet's wind blows in the direction indicated with white arrows. The atmosphere directly facing the star is hot enough to vaporize iron, and the other side is cool enough for iron to condense. Early in the transit (bottom), starlight heading toward Earth does not traverse iron vapor. Toward the end of the transit (top), starlight passes through iron vapor, producing the observed spectral absorption lines.

Spectroscopy of Smaller Planets

Transit spectroscopy is difficult even for giant planets. Smaller planets produce even weaker signals, putting them mainly out of reach of our spectrographs, for now. However, there are a few smaller planets for which transit spectroscopy has not only worked but has also led to a spectacular confirmation of a theoretical prediction: planets that orbit a star too closely are at risk of losing their atmospheres.

Gliese 436b is a Neptune-sized planet on a 2.6-day orbit around a small red star located 32 light-years away toward the constellation Leo, near the tail of the mythological lion. When the planet crosses in front of the star, it blocks 0.7% of the starlight, but when transit spectroscopy is performed, the intensity of the starlight at a wavelength of 0.122 μm is slashed by 56%. Why is more than half of the star's light blocked by a puny planet? Because the planet is surrounded by an enormous billowing plume of hydrogen gas, which has an absorption line at 0.122 μm. The dimming of the star at that wavelength lasts for much longer than the duration of the planetary transit, further demonstrating that the hydrogen cloud is much larger than the planet.

The wavelength of 0.122 μm falls in the ultraviolet range of the electromagnetic spectrum, which made these observations especially challenging. The Earth's atmosphere absorbs most ultraviolet radiation. That's good for our skin, and bad for ultraviolet astronomy. To detect the hydrogen escaping from Gliese 436b, it was necessary to use the Hubble Space Telescope, perched above the Earth's atmosphere.

Any cloud of hydrogen probably contains helium, too, because hydrogen and helium are the most abundant elements in the universe. Hot helium gas produces a dark absorption line at a wavelength of 1.083 μm, placing it in the infrared range of the spectrum. Infrared radiation at this wavelength passes through the Earth's atmosphere, but our technology for infrared spectroscopy is not as advanced as it is for visible-light spectroscopy. Only since about 2018 has it been possible to detect helium in exoplanetary atmospheres.

A good example is WASP-107b, which was studied extensively by a team led by Jessica Spake at Caltech. Imagine journeying 200 light-years toward the Virgo constellation to reach the WASP-107 system. We find a planet almost as big as Jupiter orbiting over the north and south poles of the central star every 5.7 days, a strange sight. Even stranger, when we put on infrared goggles and tune them to 1.083 μm, we see that the planet is surrounded by an expanding fog of hydrogen and helium. The fog forms a long tail trailing far behind the planet. Computer simulations of the outflowing gas suggest that the planet is losing 10 million kilograms per second (plate 9). As dramatic as that might sound, it's equivalent to losing 1% of the planet's total mass every five billion years, so the planet can afford the loss.

When we observe the escaping atmospheres of WASP-107b and Gliese 436b, we are probably seeing a slower and less extreme type of atmospheric escape than the process hypothesized to transform lower-mass planets from mini-Neptunes into super-Earths (chapter 5). The reason the escape process is slower for WASP-107b and Gliese 436b is that the planets have higher masses and stronger gravitational fields, allowing them to cling more tightly to their atmospheres.

Lava Worlds

Besides transit spectroscopy, there's another way to learn about a planet's atmosphere, and maybe even the composition of its surface. Before explaining how it works, I'd like to introduce you to Kepler-78b, a roughly Earth-sized planet discovered in 2013 by Saul Rappaport, Roberto Sanchis Ojeda, and me, when we were all at the Massachusetts Institute of Technology. One of the special things about Kepler-78b is its orbital period. By 2013, we had gotten used to the idea that planets could have orbital periods as short as a few days, but Kepler-78b still managed to surprise us by having a period of 8.5 *hours*. This means the planet is *extremely* close to the star. The planet's orbital radius is only about three times larger than the star's radius. Viewed from the dayside of Kepler-78b, the star is a blazingly bright orange disk spanning an angle 70 times larger than the Sun in our sky.

We don't know why Kepler-78b has such a tiny orbit. So close to the star, any grains of dust in the star's protoplanetary disk would have vaporized in the heat, preventing the planet formation process from ever getting started. Even worse, when the star was younger, the star's radius was *larger* than the planet's current orbital radius. This is because after a cloud of hydrogen and helium gas collapses to make a star, it takes a few million years for the young star to contract to its mature size. From Kepler-78 and similar systems, we've learned that Earth-sized planets sometimes exist where they could not have formed.

Based on its overall density, Kepler-78b is probably a solid planet composed mainly of rock and metal. Given the proximity to the star, the expected surface temperature of Kepler-78b is high enough to melt almost all common rock-forming minerals, making it likely that at least some of the planet's dayside surface is molten. It's also way too hot for the planet to have an atmosphere of ordinary gases such as nitrogen and oxygen. If there's an atmosphere at all, it's probably composed of silicate and metal vapors. Theoreticians have developed models for Kepler-78b in which a sizzling sea of lava covers a third of the planet's surface. They've also given this new category of planets an evocative name: *lava worlds*.

176

The dayside of a lava world should be glowing brightly from its own heat. And, indeed, we were able to detect light from the dayside of Kepler-78b. The trick was not to observe a transit—after all, during transits, the planet's dayside is facing *away* from us, and toward the star. Instead, the trick was to keep watching after the transit ended. Imagine watching a transit up close. Just after the transit ends, the planet's black silhouette moves away from the stellar disk. As the planet continues around its orbit, its bright dayside gradually swings into view. At first, only a thin sliver of the bright dayside is visible, making the planet look like a crescent Moon. Although we can't see the crescent of Kepler-78b from the Earth, we can nevertheless record the slight increase in the system's brightness as the planetary crescent emerges and enlarges (figure 6.7). Two hours later, when the planet has traveled one-quarter of the way around its orbit, the crescent has grown into a half-illuminated disk, like the Moon in its "first-quarter" phase. As more of the planet's dayside comes into view, the system's total brightness continues to rise. Just before the planet completes half an orbit, its dayside is almost directly facing us, and the planet is near its "full-Moon" phase. Suddenly, though, the planet passes behind the star and its light is hidden from view. The total light from the system drops. The passage of the planet behind a star—the counterpoint to a transit—is called an *occultation*.

During the Kepler mission, every time Kepler-78b was occulted by its star, the telescope registered a drop in brightness of about 10 parts per million (0.001%). This information, combined with the transit data, is enough to calculate the intensity of the light coming from the planet's dayside. From this, we estimate that the surface temperature is somewhere between 1900°C and 2600°C. The reason for the wide range of possi-

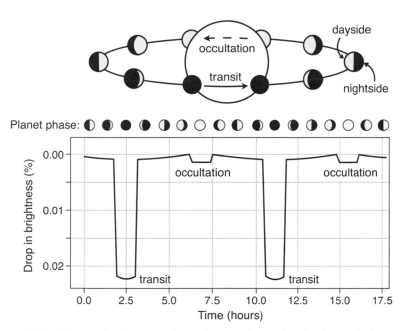

FIGURE 6.7. The changing phase of an exoplanet. The plot shows idealized brightness measurements of the Kepler-78 system. When the planet is transiting, the brightness drops by 0.022%. Afterward, the planet's glowing dayside rotates into view, causing the brightness to rise slightly. When the planet is occulted by the star, the brightness drops by 0.001%.

ble temperatures is that the light from the planet's dayside comes not only from the glow from the planet's incandescent surface, but also from reflected starlight, and we don't know whether the surface is highly reflective or relatively dull. By observing occultations over a range of wavelengths, we could in principle distinguish between the emitted and reflected radiation, because reflected radiation should have approximately the same spectrum as the star. With a good enough spectrum spanning an occultation, we might even be able to discern features in the spectrum of the planet's radiation that would inform us about the composition of the planet's surface.

Occultation spectroscopy has been performed for a few dozen hot Jupiters but only a few smaller planets. This will be another job for the Webb Space Telescope, which began scientific observations in 2022. Webb can reveal more details about exoplanet atmospheres than any previous telescope, thanks to the large size of its primary mirror, the darkness and isolation of its remote orbit (four times farther away than the Moon), and the suite of onboard cameras that are capable of extremely precise measurements.

Webb specializes in infrared observations, but is otherwise a general-purpose telescope, just as useful for studying nebulae and distant galaxies as it is for studying exoplanets. Investigators from all fields of astronomy are clamoring to use it. The Telescope Time Allocation Committee for Webb has the unpleasant job of rejecting most of the submitted proposals. Nevertheless, a big chunk of telescope time will be devoted to transit and occultation spectroscopy. The successful proposers are seeking to learn about the compositions and climates of hot Jupiters, lava worlds, and puffball planets, among other wonders. As an example of what we might learn, here is the title of one of the approved observing programs: "A Search for Signatures of Volcanism and Geodynamics on the Hot Rocky Exoplanet LHS 3844b." The team that will be attempting this feat is led by Laura Kreidberg of the Max Planck Institute for Astronomy in Heidelberg. Can you imagine being able to detect volcanoes on an exoplanet? That should be enough to convince any of the remaining planetary scientists who are still hesitating to stop worrying and learn to love exoplanets.

CHAPTER SIX

CHAPTER SEVEN

STRANGE NEW SUNS

Science fiction authors stimulate our imaginations not only by invoking exotic planets, but also by placing them around unfamiliar types of stars. The red sun of the planet Krypton was responsible for Superman's superhuman powers. Proxima Centauri and Barnard's Star are real-life red stars that have long been popular sci-fi destinations.[1] Lagash, the planet in Isaac Asimov's *Nightfall*, has a red star in the sky, accompanied by a yellow star, two white stars, and two blue stars; with so many nearby stars, the planet only experiences darkness once every 2,049 years. Arrakis, the desert planet permeated by purple worms in Frank Herbert's *Dune*, is imagined orbiting a real

---

1. Stephen Baxter's novel *Proxima*, published in 2013, describes a planet around Proxima Centauri with similar characteristics to the real planet that was discovered three years later via the Doppler technique.

star, Canopus, which is 16,000 times more luminous than the Sun.

To this point, most of the planets I've described belong to Sun-like stars. This isn't because other types of stars lack planets. It's simply because Sun-like stars have been explored more thoroughly. The focus of this chapter, though, is about the rarer sightings of planets around un-Sun-like stars.

We often hear that our Sun is a typical humdrum star, just one of a hundred billion similar stars in our galaxy. This factoid is not quite correct. The truth is that the Sun is well above average. Within 100 light-years of the Sun there are about 10,000 known stars, of which 95% are smaller and less luminous than the Sun. The statistics are probably similar throughout the galaxy. The smallest stars emit almost all their radiation in the infrared range of the spectrum, rather than visible light. At the other extreme, a small minority of stars are much more massive than the Sun and glow brightly at ultraviolet wavelengths. The galaxy is also home to a variety of more exotic stars, some that flicker in brightness like a guttering flame, and others that will eventually explode.

Why are there so many types of stars? There are two main variables that account for most of the differences between stars: they are endowed with different *masses*, and they have different *ages*.

Massive Stars

Let's start with mass. Remember that stars are born when a cloud of gas undergoes gravitational collapse. Imagine a spherical cloud of gas with a radius that gradually shrinks by a factor of a million, forming a spherical star. Now, please erase that image from your mind—it's totally unrealistic. Real gravitational collapse is a mess. Galactic gas clouds are not even remotely spherical. They're ragged and irregular, and as they collapse, they twist, fold, and break into pieces. Different regions of a cloud collapse at different rates, leading to hundreds or thousands of dense clumps. Only after the clumps have collapsed to the size of a few AU does gravity become strong enough to reshape the clumps into spheres, which ultimately become stars with a variety of masses.

The smallest possible mass of a star is about 0.08 $M_\odot$ (8% of the mass of the Sun).[2] The formation of a lower-mass object doesn't release enough heat to ignite long-lasting nuclear fusion reactions. The largest possible mass for a star is probably about 150 $M_\odot$. Above that mass, the star would generate so much heat and radiation that it would blow itself apart. Most of the known exoplanets orbit stars with masses in the relatively narrow range between 0.7 $M_\odot$ and 1.3 $M_\odot$. That's generally what we mean by *Sun-like* stars. Lower-mass stars do not emit much light, making it difficult to perform precise measurements. Higher-mass stars are very luminous but they don't cooperate when we look for planets around them using the Doppler and transit methods. If a star is more massive, a given planet causes the star to move more slowly around the center of mass, producing a weaker Doppler signal. Likewise, if a star is larger, a transiting planet blocks a smaller fraction of the star's light, making the planet harder to detect. Making matters worse, massive stars tend to pulsate and rotate quickly,

---

2.   Objects with masses between about 0.03 $M_\odot$ and 0.08 $M_\odot$ are generally classified as *brown dwarfs*, with whom we made our first acquaintance in chapter 3, and objects with masses between about 0.003 $M_\odot$ and 0.03 $M_\odot$ are in the no-man's-land separating brown dwarfs and giant planets.

producing large Doppler shifts and brightness variations that obscure the effects of any orbiting planets.

That's why we know relatively little about the planetary systems that might exist around famous bright stars such as Sirius and Vega, which are both about twice as massive as the Sun. While there have been a few amazing discoveries of planets around massive stars, most of them were not based on the Doppler or transit methods. Instead, they were found with the *direct-imaging* method. As described in chapter 3, direct imaging—the search for fireflies near searchlights—is extremely difficult and is not currently the main source of our knowledge about exoplanets. In recent years, though, the direct-imaging technique has made great strides.

The crème de la crème of the direct-imaging method is HR 8799, a bluish-white star with a mass of 1.5 $M_\odot$. In the night sky, it's located midway along one of the sides of the Great Square of Pegasus and is bright enough to see with a pair of binoculars. In 2008, a team of astronomers led by Christian Marois released an image of HR 8799 showing three faint points of light near the position of the star (plate 10). A fourth was discovered a few years later. There's no doubt that these four faint dots are planets, because enough time has passed since 2008 that the planets' orbital motion is apparent. Each planet has advanced along a slightly curved path that constitutes a small fraction of its entire orbit around the star. One of the team members, Jason Wang, made a spectacular movie of the planets that you should seek out[3] and show to your friends. Watch as four planets obey Kepler's laws around a distant star! It's suspected that the planets are all in resonance,

---

3. Try jasonwang.space/orbits.html, or search for the "HR 8799 in Motion" video. While you're at it, enjoy Wang's other exoplanet direct-imaging movies.

with orbital periods in the ratios 8 to 4 to 2 to 1. The estimated orbital periods range from about 50 to 500 years, and the orbital distances from about 16 AU to 72 AU, making the HR 8799 planets more widely spaced than the planets in the Solar System.

Marois, Wang, and their colleagues achieved this feat by using specialized cameras with two features: *adaptive optics* and *coronagraphs*. These marvels of optical engineering help astronomers fight back against a star's overpowering glare and against the ruinous blurring effect of the Earth's atmosphere.

The blurring is reduced by the adaptive optics. Once a beam of starlight is concentrated by the telescope, it bounces off a mirror whose surface can be deformed by adjusting hundreds of *actuators*—tiny devices mounted on the back of the mirror that can push and pull with precisely computer-controlled forces. The actuators allow the mirror's reflective surface to adapt to the changing conditions of the Earth's atmosphere. Every few milliseconds, the computer captures an image of a bright star, measures the distorted appearance of the star in the image, and issues commands to the actuators to alter the mirror's shape in order to focus the starlight as tightly as possible. It's a much fancier version of the image-stabilization capability of some commercially available cameras.

Meanwhile, the star's bright glare is suppressed by the coronagraph. There are many types of coronagraphs, but they all rely on the same strategy. Insert some carefully designed obstacles into the beam of starlight as it travels through the telescope, which either absorb the light coming directly from the star or cause it to diffract away from the camera—while at the same time allowing any light from nearby faint sources to pass through unaffected.

With this hardware, and sophisticated image-processing software, astronomers have detected several dozen planets. The directly imaged planets are different from the thousands of planets detected with the Doppler and transit methods. The direct-imaging planets are younger, more massive, and farther from their stars. These are precisely the characteristics that make the planets detectable with current technology. For example, HR 8799 is about 50 million years old, much younger than the 4.6-billion-year-old Sun, and the HR 8799 planets are all several times more massive than Jupiter. Such young and massive planets emit much more radiation than Jupiter. During its formation, a young giant planet attracts hydrogen and helium gas from the surrounding protoplanetary disk, causing the gas to crash into the planet at high speed and produce a lot of heat. Since the HR 8799 planets formed recently, they are still toasty hot and brightly glowing.

Why do such massive and wide-orbiting planets exist at all? The core accretion theory for giant planet formation (chapter 2) struggles to answer this question. So far from the star, the dust grains in the protoplanetary disk should have been moving too slowly for them to accumulate into a solid planet and trigger runaway gas accretion. Nobody has come up with a physically plausible way for giant planets to form closer to the star and then migrate outward to where we see them today. To solve this problem, some theorists are working on modifications of core accretion theory. Others are pursuing a completely different theory in which a giant planet is born like a star, through the gravitational collapse of a clump of hydrogen and helium gas within the protoplanetary disk.

Another showpiece of the direct-imaging method is Beta Pictoris, the second-brightest star in the southern constellation of Pictor (plate 11). Beta Pictoris a hefty young star, with an estimated mass of 1.8 $M_\odot$ and an age of about 20 million years. The combination of high mass and youth is not a coincidence. Massive stars don't shine for as long as Sun-like stars because massive stars are hotter, causing them to radiate more energy and exhaust their nuclear fuel faster than the Sun. This explains why massive stars are almost always younger than the Sun, and why they're good places to find young, toasty hot, brightly glowing planets. On the other hand, massive stars are more luminous than Sun-like stars, making it harder to detect the faint light from planets. For some massive stars, such as HR 8799 and Beta Pictoris, the advantages of young and bright planets have proven to outweigh the disadvantage of the star's brighter glare.

Even before the discovery of any exoplanets, Beta Pictoris was famous among astronomers because it was one of the first stars known to have a swirling disk of material surrounding it. The disk was discovered in 1983, not by seeing it in an image but rather by registering an abnormally high level of infrared radiation coming from the system. The excess infrared radiation is from the dust within the disk. Warmed by the star, the dust grains glow at infrared wavelengths. Soon after the initial discovery of excess infrared radiation, high-contrast imaging confirmed the disk's existence. Our view of the disk is almost exactly from the side, causing the disk to appear as a stripe of light centered on the star.

The disk around Beta Pictoris is not a *protoplanetary* disk— there's little hydrogen or helium gas left at this point. Instead, it's a *debris disk*, consisting almost exclusively of dust and chunks of solid debris left over from planet formation. A protoplanetary disk lasts for only a few million years before gradually losing its hydrogen and helium gas and becoming a debris disk, which can persist for much longer. The Solar System still has leftover dust and debris, too—not nearly as much debris as exists around Beta Pictoris, but enough to be responsible for the *zodiacal light*, a stripe of reflected sunlight that can sometimes be seen in the twilight sky.[4]

In 2010, a group of astronomers led by Anne-Marie Lagrange of Grenoble Observatory announced the discovery of a planet around Beta Pictoris based on images obtained with adaptive optics and a coronagraph. The planet, labeled "b" in plate 11, is about 10 times the mass of Jupiter and is located about 10 AU away from the star. Its orbit is nearly aligned with the plane of the debris disk, as expected, and from our vantage point the disk and the orbit are both seen almost exactly edge-on. This makes the planet travel back and forth along a straight line with a period of about 20 years although, alas, it doesn't cross directly in front of the star, so we cannot see transits. That's unfortunate, but the Beta Pictoris system has rewarded astronomers in other ways. In 2019, using a newer and more sophisticated camera, Lagrange's team spotted a second giant planet closer to the star, labeled "c" in plate 11. We don't know yet if this planet transits the star, but it's unlikely. The TESS telescopes did observe dips in the brightness of Beta Pictoris, but the dips occurred too erratically and had durations that were too long to be attributed to transiting planets. Instead, they were probably from *comets* in the Beta Pictoris system. In the Solar System, comets are kilometer-sized chunks of rock and ice left over from the epoch of planet formation. When a comet approaches the Sun, the ice vaporizes, creating a stream of gas and dust that flows away from the solid body to form a long tail—and if the comet comes close

---

4. Not to be confused with the Milky Way, a different stripe of light resulting from the combined glow of countless distant stars in our galaxy.

enough to the Earth, we can admire its magnificent tail in the night sky. Apparently, Beta Pictoris has comets sprouting tails, too, and the transiting tails produced the brightness dips observed by TESS.

One of the best things about directly imaged systems like Beta Pictoris and HR 8799 is that the planets' spectra can be observed without any of the sleights of hand involving transits or occultations that were discussed in chapter 6. The spectra that have been obtained of the HR 8799 planets are good enough to show absorption features due to water and carbon monoxide. Even better, in 2021, a team led by Jason Wang used the spectra to measure the rate at which two of the planets are rotating around their own axes, in addition to orbiting around the star. The approaching side of the spinning planet is blueshifted and the receding side is redshifted, causing the planet's absorption lines to span a broader range of wavelengths than would be observed from a stationary planet. By measuring the wavelength span of the absorption lines, Wang and his team found that one planet spins around every 10 hours, while the other takes about 15 hours, both of which are considerably shorter than the Earth's familiar 24-hour rotation period. Observations of a few other massive planets by Marta Bryan, of the University of California at Berkeley, have also revealed the planets to be rotating rapidly.

The gas giants of the Solar System spin fast, too, with Jupiter completing a full rotation every 9.9 hours and Saturn every 10.7 hours. The observed rapid rotation of gas giants agrees with theoretical expectations: the hydrogen and helium gas that is gravitationally attracted to a growing giant planet should add not only to the planet's mass but also its to angular momentum. The fun part will come when we have measured the rotation periods of dozens or more planets. We'll be able to learn more details about the gas accretion process by seeing how rotation depends on a planet's mass, composition, and age.

Low-Mass Stars

Let's turn now from the heavyweight stars to the flyweights. In the needlessly obscure jargon of astronomical spectroscopy, stars with masses between about 0.08 $M_\odot$ and 0.5 $M_\odot$ are designated *M dwarfs*.[5] I prefer the name *red dwarfs*, even though it makes some astronomers cringe because it appears more often in science fiction than in scientific journal articles. The redness of a red dwarf is a consequence of the relatively low temperature of its atmosphere in comparison to the Sun's atmosphere. The cooler temperatures cause the emitted light to be concentrated in the red end of the spectrum. In fact, the lowest-mass red dwarfs are redder than red, emitting a large fraction of their energy at infrared wavelengths. Red dwarfs are also smaller than the Sun. Together, the small sizes and low temperatures of red dwarfs cause them to be intrinsically faint. If the Sun is a campfire, a red dwarf is a birthday candle.

The small sizes and masses of red dwarfs are helpful for planet hunting. Consider the task of detecting transits of an Earth-like planet around two different stars: a twin of the Sun, and the red dwarf Proxima Centauri. For the solar twin, the

---

5. Stars are assigned letters based on the temperatures of their atmospheres. From hot to cool, the letters are O, B, A, F, G, K, M, L, and T (as I said: needlessly obscure). A *dwarf* is a star that fuses hydrogen into helium at its core (as opposed to a *giant*, to be discussed later in this chapter). The Sun is a G dwarf.

transit would reduce the star's apparent brightness by about 0.01%. Such a tiny signal can only be reliably detected with a space telescope costing hundreds of millions of dollars. Proxima Centauri is a little less than one-seventh of the Sun's size, making its cross-sectional area about 50 times smaller and allowing an Earth-sized planet to block 0.50% of its light instead of 0.01%. A brightness dip of 0.50% can be detected with a telescope and camera that you could buy for a few thousand dollars.

The advantage of red dwarfs is compounded further when trying to detect a *habitable-zone* planet. For this task, the low light level of a red dwarf makes a big difference. Just as you would need to huddle closely around a birthday candle to feel the same warmth as from a campfire, a planet needs to have a tiny orbit around a red dwarf to be heated to the same temperature as the Earth. Proxima Centauri is 600 times less luminous than the Sun, causing its habitable zone to be centered at about 0.04 AU instead of 1 AU, or in terms of orbital periods, 8 days instead of 365 days. Smaller orbits also help by increasing the geometrical probability of viewing the orbit from an appropriate direction to see transits, and when transits do occur, shorter periods help by making transits occur more frequently. It's a lot less work to monitor a star for 8 days than 365 days.

These advantages of red dwarfs must be weighed against some serious drawbacks. Their intrinsic faintness often prevents us from collecting enough light for precise measurements. In fact, red dwarfs are so faint that it took a long time for astronomers to realize that they are the galaxy's most common stars. Not a single red dwarf is bright enough to be seen by the naked eye. Even though Proxima Centauri is the nearest star to the Sun,[6] it's so inconspicuous that it wasn't discovered until 1915. The night sky gives a misleading view of the galaxy's contents for the same reason that *People* magazine gives a misleading view of the Earth's population: they are both dominated by the most atypical and attention-getting stars. Most of the stars familiar to stargazers are showoffy massive stars and giant stars, which are intrinsically rare but are luminous enough to be visible to the naked eye even when they are hundreds of light-years away.

It doesn't help, either, that a large fraction of the radiation produced by a red dwarf is in the infrared range of the spectrum. Infrared spectrographs are more expensive and difficult to build than visible-light spectrographs, and the Earth's atmosphere is not as transparent to infrared radiation as it is to visible light.

Many exoplanet-seekers have decided that the advantages of red dwarfs outweigh the drawbacks. They've built specialized instruments and even entire observatories for the sole purpose of finding planets around red dwarfs. The extraordinary TRAPPIST-1 system, with its seven planets singing in seven-part harmony, was found with a telescope built by a team led by Michaël Gillon with this purpose in mind.[7]

The designers of the Kepler mission, described in chapter 5, chose to search a few thousand red dwarf stars in addition to the Sun-like stars that made up the bulk of their target list.

---

6.   Proxima Centauri, at 4.2 light-years, is the nearest star only by a hair. It follows a wide orbit around a pair of stars, Alpha Centauri A and B. In a few hundred thousand years, Proxima will be on the far side of its orbit and the Alpha stars will take turns being closest to the Sun.

7.   Michaël is a Belgian astronomer and claims that TRAPPIST stands for TRAnsiting Planets and PlanetesImals Small Telescope and is only incidentally the name of a Belgian style of beer.

CHAPTER SEVEN

The premier discovery from the red-dwarf survey was Kepler-42, one of the very smallest stars that was monitored by the Kepler telescope during its four-year primary mission. The Kepler-42 system has three transiting planets, all of which are contained within one-sixtieth of an AU. All three planets are smaller than the Earth, with orbital periods of 11, 29, and 45 hours. In chapter 2, we shrank the Solar System by a factor of 13 billion, making the Sun into a grapefruit and the Earth a poppy seed located ten paces away. On the same scale, the Kepler-42 star would be a grape, not a grapefruit, and the three poppy-seed planets would be within 20 centimeters. The whole model would fit on a placemat.

Such tightly packed planetary systems are often found around red dwarfs. Based on analyses of Doppler and transit surveys, the probability that a star has a detectable planet with a period shorter than 100 days is three or four times larger for red dwarfs than it is for Sun-like stars. When combined with the ubiquity of red dwarfs, this fact implies that Kepler-42 is emblematic of a very common type of planetary system.

The advantages of red dwarfs extend to transit spectroscopy of their planets' atmospheres. Because of the small size of a red dwarf, the atmosphere of a transiting planet of a given size blocks a larger fraction of the star's light, which, in turn, allows us to perform transit spectroscopy of smaller planets than would otherwise be possible. This raises the exciting possibility of investigating the atmospheres of Earth-sized planets located within the habitable zones of nearby red dwarfs, which, in turn, will initiate a new phase in the search for life elsewhere in the universe. The underlying idea is that biological processes can produce detectable changes in a planet's atmosphere. Life on Earth proves the point. Oxygen is abundant in our atmosphere only because of photosynthetic organisms:

plants, algae, and cyanobacteria. These organisms harvest energy from sunlight through a complex series of chemical reactions that have the net effect of extracting carbon dioxide and water from the air and releasing oxygen. Without photosynthesis, almost all the oxygen in the atmosphere would disappear on geological timescales, because it would react with minerals in the Earth's crust and be incorporated into oxide ores. The Earth's atmosphere would become oxygen-poor, just as it was for the first few billion years of its existence, before the rise of photosynthesis.

So, even if we cannot visit exoplanets any time soon, or see their surfaces, we might be able to find indirect signs of life by looking for oxygen in their atmospheres. Scientists in the increasingly popular field of *astrobiology* are trying to establish whether the only plausible way for a rocky planet to accumulate atmospheric oxygen is through a biological process, similar to photosynthesis in spirit if not in detail. They've tried to think of other ways that oxygen gas might be produced on planets with different conditions, and how we might distinguish nonbiological processes from the effects of exoplanetary plants.[8] For example, in one nonbiological scenario, a planet is heated so strongly by the greenhouse effect that its oceans evaporate, lofting water vapor high in the atmosphere. Up there, high-energy radiation from the star splits water molecules into hydrogen and oxygen, causing a buildup of oxygen gas. We might be able to tell this is happening by observing a high abundance of water vapor along with the oxygen. Astrobiologists are also trying to think

8. Should we call plants on other planets "exoplants"? Or would that be too confusing?

CHAPTER SEVEN

of molecules besides oxygen, or combinations of molecules, that would allow for a reliable diagnosis of an atmosphere that has been altered by life.

Transit spectroscopy of Earth-sized planets inside the habitable zones of red dwarfs remains a task for the future. Part of the reason it hasn't been done yet is that we're still busy searching nearby red dwarfs for transiting planets. Proxima Centauri has a potentially rocky planet located within its habitable zone but the planet does not transit. Its orbit isn't aligned correctly. The TRAPPIST-1 system has two transiting Earth-sized planets within its habitable zone, but the star is so faint that the Hubble Space Telescope has not been able to perform useful transit spectroscopy. Even Hubble's successor, the Webb Space Telescope, might not be capable of obtaining data with the necessary precision to detect or rule out oxygen in the atmosphere of the TRAPPIST-1 planets. We may need to find suitable planets around brighter stars or build bigger space telescopes.

Searching for atmospheric "biosignature gases" such as oxygen in the atmospheres of potentially Earth-like planets is such an alluring goal that scientists and engineers have already begun planning future telescopes and instruments that will be even more capable than Webb. However, some problems with the biosignature-gas method for seeking life must be acknowledged.

One problem is the possibility of cloudy weather—not on the Earth, but on the exoplanet. To explain what I mean, consider a mini-Neptune called GJ 1214b, which orbits a red dwarf located about 50 light-years away. As a mini-Neptune, its solid surface (if it exists) is probably smothered by a hydrogen-helium atmosphere exerting a pressure thousands of times higher than the air pressure on Earth. It's not a good place to search for life as we know it. Still, because of the star's small size and proximity to the Sun, the atmosphere of GJ 1214b can be studied more easily than that of any other known exoplanet of comparable size, making it a popular target for spectroscopy. Astronomers used the Hubble Space Telescope to capture a total of about 1,200 spectra spanning the times of planetary transits—but the results were disappointing. The planet's spectrum was featureless, showing little if any variation in intensity with wavelength. Why weren't there any spectral absorption lines? The consensus is that that planet's upper atmosphere is cloudy. For successful transit spectroscopy, the planet's upper atmosphere needs to be at least partially transparent to light with wavelengths outside of the spectral absorption lines. Thick clouds block the light from the star by the same amount regardless of wavelength. Pervasive clouds may also be responsible for null results that have been obtained for several other exoplanet atmospheres. The opportunity to study exoplanetary clouds is interesting in itself—especially for hot Jupiters, where the clouds might be composed of vaporized metals and minerals—but clouds are frustrating to those of us who just want a clear view of the contents of the atmosphere.

Another problem is that some astrobiologists think a planet orbiting a red dwarf should not be considered habitable even if its surface temperature allows for liquid water. They worry that the habitable zone of a red dwarf is located too close to the star. When red dwarfs are young, they emit ultraviolet radiation, X-rays, and energetic particles that would bathe their habitable zones in dangerous radiation. Also, tidal forces are strong enough within the habitable zone of a red dwarf to synchronize a planet's rotation period with its orbital period. As a result, half the planet would permanently face the star,[9] and the contrast in temperature between the dayside and night-side might be deleterious for life. According to some climate models, water would evaporate on the hot dayside, blow over to the cold nightside, and freeze, getting permanently stuck on the nightside as solid ice. In those models, liquid water oceans don't last long.

Another potential obstacle to life is that the smallest red dwarfs glow primarily at infrared wavelengths, instead of emitting broadly across the visible spectrum like the Sun. This might create problems for photosynthetic organisms. Plants, algae, and cyanobacteria on Earth use the energy of visible light to propel the chemical reactions of photosynthesis. It's more difficult to exploit infrared radiation for chemical reactions than visible light, because each infrared photon carries less energy than a visible photon. So, it might be harder for organisms to evolve that can survive on an infrared diet.

How should we respond to these complaints about red dwarfs? Personally, I am unmoved by theoretical concerns about the difficulty for life to originate and thrive. We're too ignorant about the full range of possibilities for life for us to fret excessively over flares, tidal locking, and infrared photosynthesis. Red dwarfs are simply too convenient for planet hunting, and for transit spectroscopy, for us to ignore them.

9. Synchronization of rotation and orbital motion is also expected for hot Jupiters, lava worlds, and other close-orbiting planets. Our Moon is a local example of this phenomenon: tidal forces from the Earth slowed the Moon's rotation and its orbital speed at different rates until they matched each other. That's why we always see the same pattern of craters on the Moon.

Young Stars

Besides a star's mass, the other important variable that governs a star's appearance, size, and spectrum is the star's *age*. Stars did not form simultaneously in the remote past. Stars are forming all the time. In our galaxy, the long-term average rate of star formation is somewhere between 1 and 10 new stars per year. The youngest well-studied stars are about a million years old, the Sun is about 4.6 billion years old, and the oldest known stars are 13 billion years old, just shy of the 13.8 billion years that have elapsed since the Big Bang. One of the greatest achievements of twentieth-century astrophysics was the development and validation of the theory of stellar evolution, which explains the fascinating and counterintuitive consequences of the gradual changes that occur deep inside stars.

Despite appearances to the contrary, a star is not a peaceful and eternal source of light. A star is the location of a tense standoff in a battle between the inward force of gravity, which tries to pull all the star's mass together into a single point, and the outward force of pressure, which resists compression. The pressure arises from different physical effects depending on the type of star. For the Sun, the pressure comes from heat. Hot gas resists being compressed because of the frantic random motions and collisions of all the particles within the gas. That's why hot gas within the chambers of an internal combustion engine pushes on the pistons and makes the car go. Diving into the Sun, we would find that the temperature rises from 5,500°C to 15,000,000°C between the atmosphere and the central core. The hot gas deep within the Sun exerts more pressure than the cooler gas on the exterior, resulting in a net outward force that balances the inward pull of gravity.

This balance was achieved only after a tumultuous battle that took place much earlier. The gas cloud that collapsed to form the Sun was initially cold and diffuse, with perhaps ten atoms per cubic centimeter. The pressure of this gas was too weak to fend off gravity. The cloud collapsed toward its center of mass and formed a protostar, surrounded by a swirling protoplanetary disk. Eventually, the star's interior became dense and hot enough to ignite nuclear fusion of hydrogen into helium, which released enough energy to maintain a high interior temperature and pressure. In the resulting détente, gravity was balanced by pressure, and the rate of nuclear energy production was balanced by the rate of energy loss due to the emission of light from the star's surface.

I wish we had a time machine that would allow us to observe the formation of the Sun and the planets. Failing that, the next best thing is to search the galaxy for young systems where stars and planets are forming now. Because this action-packed phase of life is brief by astronomical standards, young planetary systems are rare, just as newborn babies are rare in the general population. And because they're rare, we need to look far away to find even the nearest examples. Their remote distances make young planetary systems difficult to study. Young stars are also intrinsically variable in brightness, hindering planet detection. Stars much younger than a million years are even harder to study because they're still enshrouded in gas and dust from the collapsing gas cloud. For all these reasons, even though we know of thousands of exoplanets in total, the number of known exoplanets younger than 50 million years can still be counted on fingers and toes.

Still, we've learned a lot from those precious few young planets, such as DS Tucanae Ab, which orbits a solar-mass star that formed approximately 45 million years ago. Observations with TESS showed that every few days, the star flares in brightness, and in between flares, the star's brightness gradually rises and falls because the star has dark blotches on its surface that move into and out of view as the star rotates. These effects made it difficult, though not impossible, to detect the transits of a planet with a radius of 5.7 $R_\oplus$ on an 8.1-day orbit. The discovery, made by a team led by Elisabeth Newton of Dartmouth College, proved that planets larger than Neptune (4 $R_\oplus$) can quickly find their way into short-period orbits. Theories that require events to play out over billions of years— for example, theories in which the gravity from a distant star gradually re-sculpts a planet's orbit—cannot easily explain how DS Tucanae Ab achieved its tight orbit in less than 45 million years.

Another fascinating system is V1298 Tauri, a young analog of the Sun with an estimated age of 23 million years. The star has four transiting planets with sizes in between those of Neptune and Jupiter and orbits that all fit within 0.3 AU (Mercury's closest distance from the Sun). Around more mature stars, we often find compact systems of multiple planets, but only rarely do we find multiple planets as large as those of V1298 Tauri. This suggests that the planets in V1298 Tauri are large *because* they're young. They might still be bloated from the heat that was released when gas accreted onto their surfaces. If this is true, then future discoveries of transiting planets over a range of ages have the potential to teach us how giant planets cool and contract with time.

The most spectacular discoveries involving young stars have come from a facility called the Atacama Large Millimeter-wave Array (ALMA), located in the Chajnantor Plateau, high in the Atacama Desert of northern Chile. Instead of visible light, which has wavelengths between 0.4 μm and 0.7 μm,

ALMA is sensitive to radiation with wavelengths between 300 μm to 3000 μm, or equivalently, 0.3–3 millimeters. For such long wavelengths, it's possible to build a type of observatory called an *interferometer* consisting of multiple telescopes that work together to make images with much finer resolution than any single telescope. ALMA has 66 shiny metal dishes that can be positioned on the plateau in different configurations within a circle of maximum diameter 16 kilometers. In its most spread-out configuration, the data from the dishes can be combined to make images that have roughly the same angular resolution as a hypothetical single radio dish with a diameter of 16 kilometers.[10] The reason ALMA was built on a dry mountain plateau is that water vapor strongly absorbs and radiates millimeter waves, spoiling the millimeter-wave view of the heavens.

Starting in 2013, ALMA has been making images of protoplanetary disks around the nearest young stars. They are breathtaking (plates 12 and 13). One of the earliest ALMA images was of HL Tauri, an infant Sun less than a million years old. The ALMA image shows a huge disk swirling around the star that extends to 100 AU. The disk is circular, but since we are viewing it from an angle of about 45°, the disk looks elliptical in projection. The glowing disk is interrupted by dark rings, the interpretation of which is unclear. According to one theory, we're witnessing a direct effect of planet formation, with each dark ring marking a range of orbital distances where the dust has been captured or flung out into space by the gravitational force from a growing planet.

10. However, a single dish of diameter 16 km would be *much* better than ALMA at detecting faint sources. With ALMA, the dishes collect only a tiny fraction of the millimeter waves that fall within the 16-km circle.

ALMA's images, amazing though they are, are not good enough to show the planets themselves. Planets are too small and too faint in comparison to the radiation from the myriad dust grains spread throughout the disk. However, in 2018, when a group of astronomers looked toward a young star called PDS 70 using a variety of different telescopes, they not only saw the star's protoplanetary disk, but also a faint point of light within the disk (plate 14). The point is 22 AU from the star and has a brightness and color consistent with expectations for a massive giant planet. Most likely, it's a baby planet. The paper reporting this tour de force had 125 co-authors. The lead author, from the Max Planck Institute for Astronomy, in Heidelberg, has the astronomically auspicious name of Miriam Keppler.

Further observations of PDS 70 have shown the planet to be glowing especially strongly at a wavelength of 0.656 μm, which you may recall (if you have a good memory for numbers) is indicative of hot hydrogen gas. This, too, is evidence of a newly born giant planet. When the gas from a protoplanetary disk is being captured by a giant planet, different streams of gas collide and crash onto the growing planet, producing shock waves and enough heat for hydrogen to glow at 0.656 μm. A second source of 0.656 μm radiation was also spotted elsewhere in the protoplanetary disk, probably marking the location of a second newborn giant planet.

All the preceding work on PDS 70 was performed not with ALMA, but with optical and infrared instruments equipped with the latest and greatest cameras and coronagraphs. When it was the mighty ALMA's turn to stare at PDS 70, the millimeter-wave image revealed a gaseous disk swirling around the second planet. This was the first-ever sighting of a *circumplanetary disk*, a disk-within-a-disk, which funnels material onto the young planet just as the protoplanetary disk funnels material onto the star. The circumplanetary disk's radius is about 1 AU, which matches the expected size of the zone within which the planet's gravity should be able to dominate over the star's gravity and accrete the gas. Strangely, though, there does not appear to be a circumplanetary disk around the first planet. Instead, there's a faint stream of dust directed toward the star, of unknown origin. We have a lot to learn, and will assuredly learn a lot, as ALMA and other instruments provide more baby pictures of stars and planets.

Giant Stars

Young stars such as DS Tucanae A and PDS 70 will eventually settle down and begin to look more like the Sun. After the planets finish forming and contracting, there will be a prolonged period of peace and quiet, as currently exists in the Solar System. During this period, the star steadily fuses hydrogen into helium at its center, releasing energy and keeping the star hot. Logically, we should call this phase of a star's existence the *hydrogen-fusing* phase, but in fact we call it the *main-sequence* phase, based on astronomical jargon from a time before nuclear fusion was understood. The main-sequence phase is the longest phase of stellar evolution, which is why almost all the known exoplanets belong to main-sequence stars. But this phase cannot last forever.

Imagine visiting the Sun five billion years from now. All the hydrogen in the Sun's core has been expended and the core is nearly pure helium, with no more nuclear fusion reactions taking place. Nuclear fusion of helium into carbon would release a lot of energy, but at this point, the core isn't hot enough to ignite helium fusion. Helium nuclei have two positively charged protons, as opposed to the single proton of hydrogen. As a result, the electrical repulsion between helium nuclei is stronger than it is between protons, and a higher temperature is required to bring the helium nuclei together closely enough to fuse. Lacking any internal source of energy to continue the standoff against the inward force of gravity, the helium core starts shrinking. Gravity seems poised to win the battle. Then, something unexpected happens.

Even though there's no more hydrogen in the core, there's plenty of hydrogen at higher altitudes. As the core contracts, the material just outside the core is pulled inward, increasing its temperature enough to ignite hydrogen fusion. So, the Sun is still generating heat through nuclear fusion—it's just that the fusion now occurs in a thin layer, or *shell*, surrounding the core. And, as it turns out, fusion in a shell is not as stable as fusion at the center of a star. As the helium core continues to shrink, the overlying shell is heated and compressed, causing the nuclear reactions to proceed more rapidly and recklessly than when the star was on the main sequence. The outcome, which is not at all obvious from first principles, is that the structure of the star changes completely. The core becomes extremely dense, and the shell of fusing hydrogen pumps so much energy into the outer layers of the star that they billow outward to many times their original size. The Sun becomes a giant star.

After another hundred million years, the Sun reaches a maximum size of about 200 times larger than its current size (figure 7.1), enveloping the current orbit of Venus. The Sun's luminosity skyrockets by a factor of 2,000. It's a fascinating sequence of events, but it also sounds like Armageddon. What will happen to the Earth? Even if the Earth avoids being en-

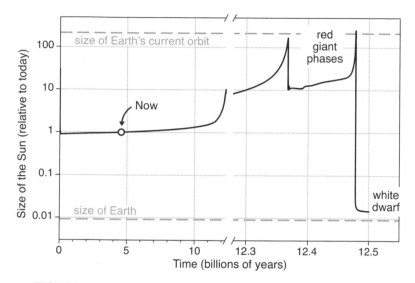

FIGURE 7.1. Theoretical calculations of the past and future size of the Sun. In about five billion years, the Sun will exhaust the hydrogen in its core and become a giant star. When the core becomes hot enough to fuse helium, the Sun will shrink. When the core helium is exhausted, the Sun will become a giant star for the second time. Finally, the Sun will shed its exterior and become a white dwarf. Note the change in the timescale between the left and right sides of the chart.

gulfed by the expanding Sun, after a while the Earth will probably spiral inward and be incinerated, as another consequence of gravitational tidal forces. As in the WASP-12 system (chapter 4), the Earth will exert tidal forces on the Sun, and the resulting distortions in the Sun's gravitational field will siphon off some of the Earth's angular momentum.

The destruction of the Earth isn't a forgone conclusion, though. Once it becomes a giant, the Sun's prodigious radiation acts like a strong wind that blows the Sun's tenuous outer layers outward into the surrounding space, and eventually out of the Solar System. As the Sun loses mass, the gravitational pull on the planets weakens, causing the planets to spiral outward into more distant orbits. This orbit-widening process counteracts the orbit-shrinking effect of tides. Neither process, mass loss nor tides, is understood well enough to predict with any confidence whether the Earth will be spared or destroyed.

Our uncertainty about the fate of the Earth is one reason why it would be interesting to observe planetary systems around giant stars. Few such planets are currently known. The transit method doesn't work because of the enormous sizes of giant stars; at the Sun's maximum future size, a transit of Earth would dim the Sun by only two parts per billion. The Doppler method works, sort of, but the precision of the measurements is hampered by the effects of pulsation and turbulent motion of the surfaces of giant stars.

Those astronomers brave enough to hunt for planets among the giants despite these obstacles have reported that wide-orbiting giant planets are just as common around giant stars as they are around main-sequence stars. One of those intrepid astronomers, Artie Hatzes of the Thuringian State Observatory in Tautenburg, Germany, led the team that bagged planets around two of the most prominent giant stars in the sky, Pollux and Aldebaran, which I mentioned in the introduction.

White Dwarfs

Even if a system of planets survives the star's growth into a giant, there are more trials and tribulations ahead. Let's return to our view of future Sun during its giant phase. The helium core is slowly contracting under its own gravity, surrounded by a shell of burning hydrogen. Compressing the core makes it hotter, and eventually, the core becomes dense and hot enough to ignite a new set of fusion reactions: helium nuclei fuse together to form carbon, and then helium and carbon fuse to make oxygen. The core is reactivated as a nuclear furnace, this time even hotter and more energetic than it was before. With a new source of heat and pressure to prevent further contraction, the core stabilizes. This causes the rest of the Sun to settle back down, too. It shrinks by a factor of 10 and settles into its new life as a helium-fusing star.

Within about 20 million years, the helium in the core is used up. Lacking a source of energy, the core begins contracting again, although this time it's made of carbon and oxygen instead of helium. A shell of burning helium surrounds the core, and at even higher elevations there's still a shell of burning hydrogen. Just as before, shell fusion causes the star to swell up. The Sun becomes a giant star for the second time, and its radiation is even more intense than in the earlier giant phase, exerting more pressure on the surrounding material and casting it off into space at a higher rate. In due course, the Sun blows away its entire exterior. We're left with a naked ball of carbon and oxygen with about half of the original mass of the Sun. Freed from the weight of all the overlying material, the core of the erstwhile Sun is exposed to the rest of the universe and is now called a *white dwarf.*

Although it is initially white-hot (hence the name), the temperature of a white dwarf is nevertheless shy of the billion degrees that would be required to start fusing the carbon and oxygen into heavier elements. Yet, despite the lack of an internal power source to replenish the energy that is radiating away into space, the white dwarf manages to resist further gravitational contraction. Over billions of years, it cools off and fades, but is never threatened by gravitational collapse. How

206

can this be, when the force of gravity is as strong as ever? Why are white dwarfs immortal?

For that, we can thank quantum degeneracy pressure, the pressure arising from the Pauli Exclusion Principle that prevents electrons from encroaching too closely on each other's territory. We encountered this peculiar source of pressure in chapter 5 when discussing the relationship between the mass and radius of giant planets and brown dwarfs. For those objects, the internal pressure is a combination of ordinary material forces and degeneracy pressure. The electrons within a white dwarf are pressed much more tightly together than in a giant planet or brown dwarf. With a mass of 0.5 $M_\odot$ and a radius of 0.01 $R_\odot$ (comparable to the radius of the Earth), the overall density of a typical white dwarf is on the order of 1,000,000 g/cm^3. At such outrageous densities, ordinary material forces pale in comparison to degeneracy pressure. White dwarfs are planet-sized demonstrations of the Pauli Exclusion Principle.

Is it possible for planets to survive the dramatic events that accompany the emergence of a white dwarf? Could there be life on a planet around a white dwarf, carrying on despite the swelling-up and eventual senescence of its parent star? Some theorists have begun speculating on these possibilities, and some astronomers have started searching for planets around white dwarfs.

For decades, there were hints that planets do exist around white dwarfs. About one-third of white dwarfs have spectra that show absorption lines of magnesium, aluminum, silicon, calcium, iron, nickel, and other heavy elements—which was unexpected, because such heavy elements shouldn't be able to remain near the surface of a white dwarf. They should sink deep into the interior and out of sight. The seemingly inescapable conclusion was that heavy elements fell onto the surfaces of the white dwarfs in the recent past. And where did those heavy elements come from? Perhaps from planets or smaller bodies such as asteroids. Collisions between these objects might produce dust that falls onto the star, vaporizes, and produces spectral absorption lines.

A breakthrough was made in 2015, when a team led by Andrew Vanderburg, then at the Harvard-Smithsonian Center for Astrophysics, detected dips in brightness of a white dwarf named WD 1145 + 017. In some respects, the dips resembled planetary transits, but their long durations, erratic timing, and other features were inconsistent with planets. Instead, the best hypothesis is that the objects are small rocky bodies that are disintegrating due to intense radiation from the white dwarf. Plumes of dust erupt from each body and spread out in the surrounding space like a comet's tail. As the chunks of rock circle around the white dwarf, their dust plumes trail behind and cover up part of the white dwarf, explaining the dips in brightness. These transiting objects might be pieces of a rocky planet that was wrecked by collisions—just the sort of collisions that are thought to be responsible for spraying white dwarfs with heavy elements.

If this interpretation is correct, it would be another gift to astronomers from Mother Nature. To help us study the compositions of rocky planets, she smashes them into dust, vaporizes the dust, throws the vapor in front of a bright light source, and transmits the resulting absorption spectra. We could not have asked for a better laboratory assistant. Spectral analysts, poring over the element-abundance patterns that white dwarfs display, have concluded that most of the falling material has heavy elements in the same proportions as the Earth. A few of them have higher proportions of lighter and more volatile elements such as oxygen, carbon, and nitrogen. The higher abundances of volatile elements might be from chunks of ice falling onto the star—perhaps from "water worlds" similar to the icy moons of Jupiter—in addition to the rocky material.

To support this interpretation, it would be nice to find intact planets around a white dwarf in addition to the dust and debris. In 2020, Vanderburg and his colleagues scored again by using TESS data to detect the transits of a Jupiter-sized object, presumably a planet, around the white dwarf WD 1856 + 534. Like most white dwarfs, this one is not much bigger than the Earth. So, in a strange reversal, the planet in this system is seven times larger than the star. To make an elementary school model, we would use a grapefruit for the *planet* and a grape tomato for the *star*.

Does WD 1856 + 534 give us a glimpse of the Sun's future, eight billion years from now? Maybe the system's inner planets were destroyed when the star became a giant, and the outer planets survived. Could one of the outer planets have spiraled inward due to tides, accounting for the transits Vanderburg discovered? It's even conceivable that the giant planet was engulfed by the star's expanding outer atmosphere, and spiraled inward while it was within the star itself. At this point, the discovery raises more questions than it answers.

Exploding Stars

Sun-like stars end up as white dwarfs. Lower-mass stars, such as red dwarfs, will in principle become white dwarfs, but red dwarfs sip their hydrogen so slowly that the universe is not yet old enough for any of them to have become white dwarfs.

CHAPTER SEVEN

High-mass stars, on the other hand, have a different fate. Their planets are in for a shock.

Consider, for example, the true story of a 20 $M_\odot$ star named Sanduleak −69 202. It shone brightly for 10 million years before converting all the hydrogen at its core into helium. Because of the star's large mass, the core was also hot enough to fuse helium into carbon and oxygen. Unlike the case of the future Sun, though, the core of this star was too massive for degeneracy pressure to halt further collapse. The core continued contracting and heating, causing carbon and oxygen to fuse into heavier elements such as neon, sodium, and magnesium.

You might think this could go on forever, with heat being generated by successive rounds of fusion into heavier elements. Alas, this is not so. Fusion of progressively heavier elements releases less and less energy. The effort to oppose gravity became increasingly desperate. Silicon and sulfur were forged from oxygen, but they lasted only a few days before being transmuted into iron and nickel, and that was the end of the line. Iron and nickel are the elements with the most stable possible nuclei. If you try to fuse them by squeezing them together, you are not rewarded with energy; instead, the reactions absorb energy.

Suddenly, the battle between fusion-powered pressure and gravity ended, with gravity the winner. Acting unopposed, gravity needed only a few tenths of a second to shrink the star's core from a radius of 10,000 to 50 kilometers. The electrons and protons were compressed so tightly that most of them merged to make neutrons. With neutrons and protons in nearly direct contact with each other, further contraction was prevented by the *strong nuclear force*, the fundamental force of nature that gives atomic nuclei their rigidity.

It sounds like a happy ending, with another source of pressure becoming available just in time to prevent gravitational collapse, but the sudden entrance of the strong nuclear force had more dramatic consequences. Because of the enormous strength of the strong nuclear force, the star's core went from being compressible to being extremely rigid within a very short time interval. Material falling from further above slammed into the rigid core and was reflected backward, launching an outward shock wave. The implosion reversed and became an explosion: a supernova.[11] The shock wave tore through the overlying material, producing a fireball of a scale almost beyond description. The total amount of energy released in the explosion exceeded the Sun's lifetime energy output over 4.6 billion years.

None of the events taking place within the star were observed directly—my description is based on theoretical calculations of what probably happened—but the fireball was sighted on Earth on February 24, 1987, as a brightening point of light visible to the naked eye for several months. The new light appeared in a nearby galaxy called the Large Magellanic Cloud. This was Supernova 1987A, the nearest known supernova explosion in nearly four centuries. All the fireworks served as the obituary for the star Sanduleak −69 202, and the birth announcement for the neutron star at its center.

The violence of this story helps to explain why astronomers nearly fainted from surprise in 1992, when Alex Wolszczan and Dale Frail found a planetary system around a different and much older neutron star. Any planets should have been oblit-

11. To be specific, this is a *core-collapse* supernova. There are other types of supernovae that do not produce neutron stars and are not as well understood.

CHAPTER SEVEN

erated by the explosion that accompanied the birth of the neutron star. Nevertheless, as described in chapter 3, Wolszczan and Frail took advantage of the radio pulses coming from the neutron star to discover two planets, and Wolszczan later went on to discover a third. The origin of these planets is unknown. They probably formed from the detritus of the supernova explosion, near the beginning of the star's afterlife as a neutron star. The International Astronomical Union invites the general public to nominate nicknames for exoplanets; for the pulsar planets, the chosen nicknames were Poltergeist, Phobetor, and Draugr, mythological references to ghosts, nightmares, and the undead.

The pulsar timing data are precise enough to reveal the effects of the gravitational forces between the planets, providing unusually detailed information about the system. In this respect, the pulsar timing anomalies resemble the transit-timing variations described in chapter 6. Two of the pulsar planets have masses of about 4 $M_\oplus$ and are engaged in a perfect-fifth resonance, with the inner planet revolving three times for every two revolutions of the outer planet. Their orbits are contained within 0.5 AU, are aligned to within 10°, and are nearly circular, with eccentricities smaller than 0.03. The third planet's orbit is closer to the neutron star, and its mass is only 0.02 $M_\oplus$, not much larger than the Moon's mass.

Because there are so few known pulsar planets, and none are as well-studied as PSR 1257 + 12, it's difficult to assess the significance of the pulsar planets and determine whether they have anything to do with normal planets. Dale Frail once compared himself to an anthropologist who made first contact with an isolated tribe with unusual customs. It's also hard to imagine a less hospitable place for life than the pulsar planets.

STRANGE NEW SUNS

Instead of a warm and friendly Sun, there is a pinpoint in the sky that spews a maelstrom of radio waves, X-rays, and lethal doses of radiation.[12]

Stars with initial masses between about 8 $M_\odot$ and 20 $M_\odot$ are expected to explode as supernovae. The neutron stars they leave behind have masses between about 1.4 $M_\odot$ and 3 $M_\odot$, and contract to a final radius of about 10 kilometers, giving an average density on the order of 10^{15} g/cm^3. This is an unfathomably high density, but if you want to attempt to fathom it anyway, try imagining the density of a 100,000-ton cargo ship compressed to the volume of a poppy seed.[13]

Stars that are initially even more massive than 20 $M_\odot$ have cores that are too massive to become a neutron star. Even the rigidity provided by the strong nuclear force cannot oppose the gravity of such a massive core. Instead, the strong force gives way, and the core collapses all the way down to become a black hole. Gravity is victorious.

If we take Einstein's theory of general relativity literally, all that will remain of such a star is a point in space occupying zero volume and endowed with the mass of several Suns. Where there once was a star, there is now a puncture in the fabric of spacetime. Surrounding the black hole is a mathematical boundary called the event horizon, within which the gravity is so intense that nothing, not even light, can avoid falling into the black hole and being obliterated. The radius of the event horizon is about 15 kilometers for

12. This has not stopped science fiction authors from dreaming about life around a neutron star; see Robert Forward's *Dragon's Egg* or Stephen Baxter's *Flux*.
13. I'll bet you weren't expecting a book about planets to feature so many poppy seeds.

CHAPTER SEVEN

a 5 $M_\odot$ black hole, and it grows in proportion to the black hole's mass.

Recent years have seen a resurgence in black hole research. Black holes have been "seen" merging with one another, or to be more precise, scientists have detected bursts of gravitational waves that propagate outward from such events, traveling through the fabric of space like ripples across the surface of a pond. There is now nearly incontrovertible evidence for a black hole at the center of our own galaxy. The immediate environments of our galaxy's black hole and the black hole at the center of a nearby galaxy called M87 have been mapped out with a radio-wave interferometer called the Event Horizon Telescope.

Planets around black holes are irresistible destinations for science fiction authors, such as the screenwriters of the 2015 film *Interstellar*, in which astronauts land on a planet orbiting a black hole named Gargantua. The strong tidal forces from the black hole raise mile-high waves in the planet's oceans. The black hole also warps the surrounding spacetime to such a degree that for every hour that passes on the planet, seven years elapse back home on Earth. The planet is heated to a cozy temperature by the warmth of a nearby star that also orbits the black hole.

I'd love to know if some black holes really do have planetary systems. The existence of planets around a neutron star makes it seem possible that they also exist around black holes. After all, a neutron star is just one sneeze away from collapsing and becoming a black hole. Unfortunately, the usual planet-hunting techniques don't work for black holes. Black holes don't emit pulses of radiation, and they have no surfaces where a lighthouse-like beam could be anchored. Black holes emit no light that could be blocked by transiting planets, and no spectra that could be checked for Doppler shifts.[14] There's an up-and-coming planet detection method called gravitational lensing that works for black holes, in principle—as described in the next chapter—but, for now, planets with black Suns are seen only in our imaginations.

14.   Some black holes appear to emit X-rays and other types of radiation, but the radiation is coming from a disk of material swirling just outside the event horizon.

CHAPTER SEVEN

CHAPTER EIGHT

THE WORLDS TO COME

Over the last few decades, exoplanetary science descended from the realm of airy speculation to become a data-driven field with a rapidly advancing frontier. Every month, someone reports a startling new planetary system, describes a creative new application of existing data, or proposes an ambitious new instrument to perform ever-more-penetrating observations. Trying to keep up with the field leaves me breathless, and the history of surprises makes me wary of offering predictions. Nevertheless, this last chapter is devoted to the discoveries to look out for in the years ahead.

For purposes of prediction, it helps to remember that progress is usually driven by improvements in technology. Whenever we improve our ability to measure the properties of starlight, we can find new types of planets. With that in mind, let's start by examining the Doppler and transit methods, the most prolific planet-finding techniques to this point. Can we expect major technological advances that will bring more discoveries? The Kepler mission brought us right to the threshold of finding Earth-sized planets in Earth-like orbits around Sun-like stars. Will Kepler's successors be capable of carrying us through?

The TESS spacecraft (chapter 5) will dutifully scan the sky for as long as the equipment keeps working and as long as NASA is willing to fund the team of scientists and engineers that look after it. Although the spacecraft has a limited supply of fuel on board for maintaining its orbit, at the current rate of usage the fuel will last for decades. With each year that passes, TESS will be able to find planets with longer orbital periods. Still, because TESS typically only observes a given star for a month at a time—a strategy that was chosen to allow the entire sky to be scanned every few years—TESS is not very good at hunting for Earth-twin planets with one-year periods. Easier prey for TESS is the population of habitable-zone planets around red dwarf stars, which have typical orbital periods of a few weeks instead of a year.

Habitable-zone planets will be studied with the Webb Space Telescope, which is capable of major advances in transit spectroscopy (chapter 6). Although it may be a stretch for Webb to detect biosignatures such as oxygen in the atmospheres of Earth-sized planets, Webb is already performing deep dives into the atmospheres of larger planets (see, for example, figure 6.5). Webb will also be able to record the spectra of protoplanetary disks with enough detail to measure their composition and check on the existence and location of the "snow line" that plays such a prominent role in planet formation theory (chapter 2). Yet another of Webb's superpowers is a coronagraph that will search the nearest stars for wide-orbiting giant planets.

CHAPTER EIGHT

Knowing that the Webb Space Telescope will always be oversubscribed, and that it will not be possible to observe all the interesting exoplanets, the European Space Agency is building a space telescope called the Atmospheric Remote-sensing Infrared Exoplanet Large-survey (ARIEL). Although ARIEL will be smaller than Webb, it will be devoted almost entirely to transit spectroscopy and should be able to observe about a thousand exoplanets. According to the current plan, ARIEL will be launched in 2029.

The European Space Agency is also planning a successor to TESS, called PLATO,[1] which will pick up the pace of planet discovery. TESS has four telescopes; PLATO will have 26. PLATO will also be able to observe continuously for longer intervals of time, giving it better odds of detecting Earth-like planets around Sun-like stars. PLATO is currently scheduled for launch in 2026. The Chinese Academy of Sciences is evaluating a proposed project called Earth Two, which would resume monitoring the same stars that Kepler monitored from 2009 to 2013. By combining data from Kepler and Earth Two, it might be possible to detect a large sample of transiting Earth-like planets around Sun-like stars, thereby finishing the job for which NASA's Kepler mission was originally designed.

The Doppler Method

When prognosticating the future of Doppler observations, the number to remember is 9 centimeters per second, the speed with which a Sun-like star would orbit around the center of

---

1. This is one of those contrived pseudo-acronyms: PLAnetary Transits and Oscillations of stars.

mass if it were being pulled by an Earth-mass planet orbiting at 1 AU. Current spectrographs are limited to a Doppler precision of about 50–100 cm/sec. Representing the state of the art is a spectrograph called ESPRESSO,[2] which has proven capable of detecting Doppler signals with an amplitude of only 40 cm/sec. In 2021, a team led by João Faria, at the University of Porto, used ESPRESSO to confirm the existence of a planet around the red dwarf Proxima Centauri. The planet has a minimum mass of 1.1 $M_\oplus$ and an 11-day period, placing it within the star's habitable zone. Faria's team also found evidence for a second planet with a 5-day period and a minimum mass of 0.3 $M_\oplus$. Today's Doppler measurements don't get much better than that. Still, the measurement precision will need to be improved by an order of magnitude to find Earth-like planets around Sun-like stars, which have longer orbital periods and produce smaller Doppler shifts. This is the problem that keeps the Doppler spectroscopists lying awake in bed all day.[3]

One of their worries is maintaining the stability of the spectrograph. The goal is to detect tiny shifts in the wavelengths of the star's spectral absorption lines. In chapter 5, I described the methods astronomers use to distinguish Doppler shifts from any slight changes in the position or temperature of the spectrograph. To get to the next level of precision, a device called a *laser frequency comb* will help. Laser frequency combs were invented by atomic physicists, not astrophysicists, and the idea is clever enough to have earned John Hall and Theodor Hänsch one-half of the 2005 Nobel Prize in Physics.

2. And another: the Echelle SPectrograph for Rocky Exoplanets and Stable Spectroscopic Observations.
3. They work at night.

CHAPTER EIGHT

They figured out how to coax a laser to emit ultra-short pulses of light at extremely regular intervals—say, every billionth of a second. The spectrum of such a rapidly pulsing light source does not resemble a continuous rainbow, but rather a series of bright lines at regular intervals, like the teeth of a comb. When astronomers merge the light from a laser frequency comb with light from a star, the spectrum shows the star's absorption lines along with the laser lines. The laser lines act as evenly spaced tick marks, ideal for measuring the wavelengths of the star's absorption lines.

In laboratory experiments, laser frequency combs are stable enough to allow velocities to be measured to within a few *millimeters* per second, more than sufficient for detecting Earth twins. Laser frequency combs have also been strapped onto actual telescopes. They're finicky instruments. In the initial deployments, it was necessary to send a team of atomic physicists to the telescope in addition to the astronomers. Further experience has allowed them to be operated with greater regularity. Although the results have not yet been as impressive as in the laboratory, laser frequency combs are such an elegant and promising solution that they are becoming standard equipment for planet hunting.

However, even after laser frequency combs become routine and reliable, we'll need to overcome another problem. Often, the precision of our Doppler measurements is limited not by the capabilities of our instruments, but by the stars themselves. The surfaces of stars are not perfectly still. In Sun-like stars, heat is transported from the interior to the outermost layers by convection, as in a pot of boiling water. Bubbles of hot gas emerge at the surface, and cooler material sinks down. The turbulent motion causes Doppler shifts that combine with (and sometimes overwhelm) any Doppler shifts due to orbiting planets. Doppler shifts can also be produced by the irregular dark spots that blemish Sun-like stars. The turbulent motion and patchy brightness fluctuations of the surfaces of stars are collectively called "stellar activity." It sounds like a hopeless situation because the stars are outside our control. The saving grace is that Doppler shifts produced by planets should repeat exactly with every planetary orbit, while stellar activity is more random. So, by observing for a long enough time span, we might be able to tell the difference. Another approach is to look for evidence of changes in the spectrum besides the Doppler shift that might allow us to diagnose stellar activity and make appropriate corrections.

The Astrometric Method

While astronomers are striving toward ever higher levels of perfection with the transit and Doppler techniques, two other planet-finding techniques will reach maturity: astrometry and gravitational lensing. Though they are based on different physical principles, these two methods share the ability to explore the *outer* regions of planetary systems. Most of the known transiting planets have orbits smaller than 0.2 AU, and most of the known Doppler planets have orbits smaller than 2 AU. With astrometry and gravitational lensing, we can look forward to discovering thousands of planets with orbits between 2 AU and 20 AU—a range of distances that, in the Solar System, encompasses Jupiter, Saturn, and Uranus.

Just as photometry is the measurement of a star's brightness, and spectroscopy is the measurement of its spectrum, astrometry is the measurement of its position in the sky. As described in chapter 3, by monitoring the latitude and longitude of a star on the imaginary celestial sphere, we can try to detect the star's motion around the center of mass of a planetary system. It's the same principle as the Doppler method, except that instead of detecting the star's speed toward or away from us, astrometry is based on detecting the star's sideways motion. The Doppler method works best for planets with small orbits and short orbital periods, because such planets cause the star to move at greater speeds. In contrast, the astrometric method works best for planets with wide orbits, because it is based on sensing the *distance* between the star and the center of mass, not the star's *speed*. Increasing the orbital distance of a planet causes the center of mass to shift further from the star.

If aliens in the Proxima Centauri system were applying the astrometric method, they would see the Sun as a point of light that traces out a rosette-like pattern as it is pulled around by all the planets, with the biggest effect coming from Jupiter (refer back to figure 3.3). The maximum angular shift would be about one-millionth of a degree, the same shift we would observe if a firefly located 1,000 kilometers away were to fly rightward or leftward—and that's for aliens from Proxima Centauri, the *nearest star*. Anyone looking from farther away would see an even tinier shift, declining in proportion to distance from the Sun.

Astrometry was the earliest measurement technique in the history of astronomy. Long before it was possible to make quantitative measurements of a star's brightness or spectrum, sky-watchers recorded the positions of stars and the planets in the Solar System. Astrometry allowed seventeenth-century astronomers to discover the laws of planetary motion and nineteenth-century astronomers to measure distances to stars using parallax, as described in the introduction. Astrometry was also the basis of the earliest announcements of exoplanet discoveries. In 1855, William Jacob, the director of Madras Observatory in India, reported a possible planet around one of the two stars in the 70 Ophiuchi system, based on the apparent wobbling motion of one of the stars. In 1969, American astronomer Pieter Van de Kamp concluded that Barnard's Star was moving in response to two giant planets. In both cases, as well as many others, the apparent shifts in position proved to be measurement errors. Measuring tiny positional shifts turned out to be harder than measuring correspondingly tiny Doppler shifts, because the Earth's atmosphere corrupts positional measurements more than it does spectroscopic measurements.

Despite these false starts, the astrometric technique has begun to blossom thanks to a European space telescope called Gaia, which is measuring the positions of stars with unprecedented precision. Gaia was not intended to be a planet hunter. Its main job is to improve maps of the Milky Way galaxy by measuring the positions, distances, and velocities of stars as far away as 25,000 light-years. For the nearest and brightest stars, the data from Gaia are also good enough to detect giant planets. Unlike the Doppler technique, the astrometric technique reveals a planet's true mass without any ambiguity. This is because measuring the two-dimensional motion of the star in the sky provides more information than the Doppler method, which is sensitive only to motion along the line of sight. In fact, when Gaia finds multiple planets around the same star, the data can be used to determine the angles between their orbital planes, which is difficult to determine in any other way.

Gaia is searching for planets around the same stars that have already been searched with the Doppler and transit techniques, allowing us to extend our knowledge about planetary systems from their inner regions to their outer regions. For example, there are hints from Doppler surveys that compact systems of inner planets tend to be accompanied by outer giant planets, which could be a clue to their formation. Gaia will provide more definitive data. Although Gaia was launched into space in 2013, it took until 2022 to announce the first batch of planet detections, because of the long time required for data collection and processing. It wasn't easy to wait patiently. Hundreds or even thousands of giant planets are expected to be discovered with Gaia data in the years to come.

The Gravitational Lensing Method

Gravitational lensing is the other up-and-coming technique that specializes in wide-orbiting planets. It's also the most exotic planet-finding method. It doesn't involve tracking a star's position, brightness, or spectrum—in fact, it's unnecessary to detect any light whatsoever from the star. Furthermore, gravitational lensing doesn't rely on Kepler's laws of planetary motion or Newton's laws of motion and gravity. It relies on Einstein's general theory of relativity.

Einstein taught us that we should not think of gravity as a force of attraction between material bodies. Gravity is actually an aspect of geometry. Space and time are components of a single entity called spacetime, and gravity is the curvature of spacetime. A star causes the surrounding space to become curved, just as a bowling ball on a trampoline puts a curved dent into the trampoline's surface. Then, when something else passes by—whether it is a tennis ball rolling on the trampoline or a beam of light passing the star—its trajectory follows the curvature of the underlying space. The curve in the trajectory is said to be the result of the "force of gravity," in the older and not-quite-correct language of Newtonian physics.

Tempted though I am, I will not try to explain how to visualize the curvature of a three-dimensional space, what it means for *time* to be curved along with space, or why a light beam is deflected even though it has no mass. Instead, we will deal with general relativity on a "need to know" basis. For the limited purpose of understanding how the theory of general relativity can be exploited to find planets, we will rely on a close analogy between curved spacetime and the optics of curved glass lenses.

The *index of refraction* of a piece of glass is the factor by which it slows down a beam of light, which in turn determines how much a beam of light is deflected when it enters the glass at an angle. Typical indices of refraction are between 1.5 and 1.7, depending on the density of the glass. Other substances have different indices; for example, water has an index of 1.3. Empty space has an index of 1, by definition. In general relativity, a star's gravitational field acts as though it raises the index of refraction of empty space, by an amount proportional to the star's mass and inversely proportional to the distance from the star. For light rays passing close to the Sun, for example, the effective index of refraction is raised from 1 to 1.0000042. This is a tiny effect but it has observable consequences, such as shifts in the apparent positions of distant stars—shifts that were successfully detected during total solar eclipses in the early twentieth century.

It's an important and subtle point that the speed of light never actually changes. An observer positioned anywhere along the curved trajectory of a beam of light will see the light rushing by at 299,792,458 meters per second, the same speed as always. What really happens is that a star's gravitational field curves the surrounding spacetime. However, if all you want to do is calculate the trajectories of light beams that pass close to a star, as viewed from afar, then mathematically you are allowed to forget about curved spacetime and pretend the star is surrounded by a sphere of glass. The glass is densest near the star and becomes less dense with increasing distance from the center. When looking through such a weird piece of glass, everything in the background looks distorted and magnified.[4]

The bending of light beams by gravitational fields is called *gravitational lensing*. I'm not fond of that name, though. A real lens is designed to bring light rays to a sharp focus or cause light rays to diverge away from a focal point. In a microscope or telescope, lenses are used to make images that are magnified but otherwise preserve the true appearance of the object being viewed. Gravitational lenses don't have focal points, nor do they preserve true appearances. The "image" formed by the gravitational field of a star is stretched and distorted with aberrations that would give a lens maker nightmares. Gravitational lenses are more like fun-house mirrors—the ones that make you look unnaturally tall or short, or that make legs sprout up and down from your waist. The French term, *mirage gravitationnel*, is better. You've probably seen a mirage while driving on a long highway on a hot day. Shallow pools of water seem to appear on the road ahead, and then vanish as you approach. The explanation is that the heat from the road causes the overlying air to expand slightly, lowering its index of refraction. This, in turn, causes light from the sky that would ordinarily hit the ground to be redirected to your eyes, making the road look bright and shiny. A gravitational mirage is

---

4. You can simulate the effect by looking at a candle through the curved bottom of a wine glass. The varying thickness of the tapered glass bottom has a similar effect as a varying index of refraction.

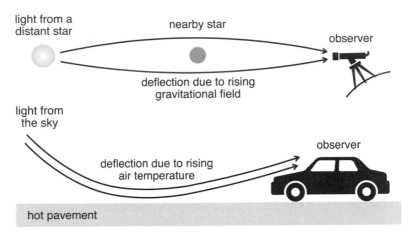

FIGURE 8.1. Formation of a road mirage (bottom) and a gravitational mirage (top). In both cases, light is deflected away from a straight-line trajectory, creating optical illusions.

similar except that the index of refraction varies due to gravity instead of temperature (figure 8.1).

In 1936, Einstein considered what would happen if two unrelated stars happened to line up on the sky, with one star in the foreground and the other star in the background. He realized that the gravity of the foreground star would create a specific type of mirage: the background star would be split into two stretched-out images. However, Einstein concluded it was unlikely that such a double image would ever be observed. The first problem is that chance alignments of unrelated stars are rare. At any moment, only about one out of every ten million stars in the sky is creating a mirage of a star in the background. The second problem is that even if you knew where to look, you wouldn't be able to see the mirage, because the two images of the background star would be too closely spaced to be distinguished by any realistic telescope. Einstein had been prompted to perform this calculation by an enthusiastic amateur scientist named Rudi Mandl. The tone of Einstein's paper is apologetic, stating, in effect, "This calculation is amusing but unimportant, and I'm only publishing it because Mandl kept bugging me about it." Einstein's brief publication turned out to be important, though. It inspired other prominent scientists to think about the gravitational deflection of light, which has since led to major scientific achievements, of which finding exoplanets is only one example.

What Einstein didn't seem to realize is that a gravitational mirage can be detected even though the two images of the background star are too close together to be distinguished. The trick is to take advantage of the overall magnification— the total brightness of the two distorted images exceeds the star's ordinary brightness—and the relative motion of the foreground and background stars. Since the stars are unrelated, they follow different paths within the galaxy, and their nearly perfect alignment in the sky is only temporary. The background star gets brighter when the foreground star drifts in front, and then goes back to normal brightness. The characteristic changes in brightness over time allow you to recognize that a gravitational mirage has just formed and dissipated (figure 8.2).

The low probability of chance alignments can be overcome by monitoring the brightness of hundreds of millions of stars at once. It's like the transit method, except that instead of looking for stars that fade, you look for stars that brighten. In 1993, the first detection of the brightening of a star due to gravitational lensing was achieved by a group led by Charles Alcock, then at the Lawrence Livermore National Laboratory. He and his team were not trying to check on Einstein's theory, nor to find planets. They were looking for brown dwarfs,

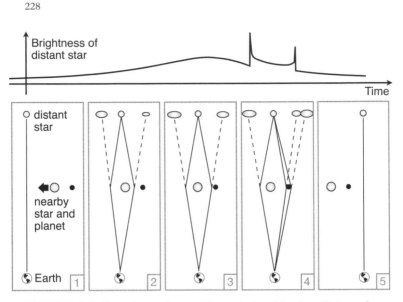

FIGURE 8.2. Detecting a planet with gravitational lensing. (1) A nearby star and its planet drift across the line of sight to a distant star. (2) The nearby star's gravity deflects the light from the distant star, producing two distorted images and an overall magnification. (3) The magnification increases as the nearby star moves directly in front of the distant star. (4) The planet's gravity creates a third image of the distant star and a burst of extra magnification. (5) Everything returns to normal when the two stars move out of alignment. Adapted from a figure by D. Bennett, NASA/GSFC.

black holes, and other dark objects roaming around the galaxy. With gravitational lensing, the only relevant property of the foreground object is its mass. It doesn't matter if the foreground object is luminous or black. This makes gravitational lensing a unique tool for finding massive objects that emit little or no light.

It's also why gravitational lensing can be used to find a planet even without detecting any light from the star it orbits. Imagine a nearby star drifting across the line of sight to a more distant star. The gravity of the nearby star temporarily magnifies the background star, causing it to rise and fall in brightness over the course of a few weeks. If the nearby star has a planet that also drifts in front of the distant star, the planet's gravity adds to the magnification, causing an extra blip of brightness lasting a day or so. The planet acts like a denser pebble of glass embedded within the fictitious glass sphere surrounding the star.

Gravitational lensing is a beautiful and fascinating technique with strengths and weaknesses relative to the other planet-finding techniques. On the plus side, the lensing signals are sometimes obvious. The extra magnification due to a planet can be as high as 10%, a booming signal in comparison to the typical signals that are sought after with the transit, Doppler, direct-imaging, and astrometric methods. Another attractive feature of gravitational lensing is greater sensitivity to planets with orbital distances of a few AU. On the minus side, gravitational lensing events are one-time-only performances. Once the stars drift apart, the show's over. With the other methods, the signal keeps repeating as the planet completes orbit after orbit, allowing for improvements in the precision of measurements and for follow-up observations with new instruments. For this reason, the gravitational lensing technique is best for collecting planet statistics—conducting a census—rather than studying individual planets in detail.

The amazing ability of gravitational lensing to detect planets independently of the light from the foreground star is both a blessing and a curse. We can use this technique to find planets around low-luminosity objects, including red dwarfs, white dwarfs, brown dwarfs, and even black holes, in principle. On the other hand, this same feature is a bug, because most of the time we only have a vague idea of what kind of object the planet is orbiting. The usual assumption is that the star is a red dwarf, simply because red dwarfs are so numerous. One way around this problem is to wait a long time for the foreground star to move away from the background star and try to detect its light with a large telescope.

The first discovery of a planet using gravitational lensing was in 2003, and since then, more than a hundred planets have been discovered with this method. The results of a gravitational lensing census cannot be stated accurately and concisely in nontechnical terms, partly because the stars in the census span a wide range of masses and distances from the Sun, but here's the gist of the most important finding so far: the most common type of planet with an orbital distance between 1 AU and 10 AU has a mass comparable to that of Uranus and Neptune (about 15 $M_\oplus$), rather than Saturn and Jupiter (100–300 $M_\oplus$). This discovery created yet another problem for the core-accretion theory of giant planet formation. According to the theory, giant planet formation is a threshold process: if a solid planet grows more massive than about 8 $M_\oplus$, it should quickly accrete the surrounding gas in the protoplanetary disk and become a fully fledged giant planet like Saturn and Jupiter. Planets of intermediate masses, between 10 $M_\oplus$ and 100 $M_\oplus$, were predicted to be rare, which is why it's interesting that such planets are the *most common* type of planet found through gravitational lensing.

The gravitational lensing technique is poised to make greater advances. NASA has begun construction of a new space observatory called the Roman Space Telescope,[5] which is tentatively scheduled for launch in 2027. Roman will specialize in infrared observations and will be able to make images over a

5.   Named after Nancy Grace Roman, NASA's first chief astronomer, who played a major role in getting the Hubble Space Telescope off the ground.

CHAPTER EIGHT

field of view approximately ten times wider than those of the Hubble or Webb telescopes. Roman's original purpose was unrelated to exoplanets. It was envisioned as a telescope for cosmology, the study of the universe as a whole rather than individual planets, stars, or galaxies. Cosmologists want to survey the entire observable universe, including very distant galaxies. Infrared observations are essential because the visible light from distant galaxies is Doppler-shifted into the infrared range of the spectrum by the expansion of the universe. Wide-field observations are also essential to measure the properties of millions of galaxies.

That's why cosmologists wanted the Roman telescope, but it didn't take long for exoplanet scientists to realize that Roman would also be well equipped to find planets with the gravitational lensing method. By observing at infrared wavelengths, it will have a less obstructed view of the Milky Way. In visible light, the Milky Way looks like a long stream of spilled milk— hence the name—but it is interrupted by dark blotches in several locations, as if a sponge has dabbed up the milk. Those dark patches are clouds of gas and dust in the foreground. If your eyes could see near-infrared radiation, you'd see the profusion of distant stars in those patches, because dust doesn't block infrared radiation as much as visible light. By penetrating the dust, Roman will gain access to more stars and improve the odds of finding the rare cases of gravitational lensing.

The plan for the Roman telescope is to split the observing time between cosmology and exoplanets. During the exoplanet portion, it will spend a year staring at the center of the Milky Way, keeping tabs on a hundred million faraway stars to identify "lensing events," when stars brighten due to gravitational lensing by objects in the foreground. Roman should

THE WORLDS TO COME

detect planets over a wide range of masses and orbital distances, including Earth-like planets, and analogs of all the other Solar System planets except Mercury, whose orbital distance and mass are too small to produce noticeable gravitational lensing effects.

Roman will also be capable of detecting any *free-floating planets* that are drifting by themselves throughout the galaxy, untethered to any star. Recall from chapter 4 that some of the theories for the unusual architectures of planetary systems involve close encounters between giant planets and orbital rearrangements. Such encounters would occasionally cause planets to be ejected into deep space. A free-floating planet can be detected not by looking for an extra blip of brightness during a lensing event, but rather, by looking for lensing events with durations too short to be caused by a star. All else being equal, the duration of a gravitational lensing event varies in proportion to the square root of the mass of the foreground object. When a star acts as the gravitational lens, the brightness enhancement lasts several weeks, but when an isolated planet is the lens, the typical duration is shorter than a day. There have already been claimed detections of free-floating planets, but there has been controversy over the interpretation of the data, with arguments over whether the signals are genuine, and if so, whether they arise from free-floating planets or planets with wide orbits around stars. If castaway planets exist, as they should, then Roman will be able to detect them more securely.

The Roman telescope is not only a great scientific project—an unexpected synergy between two different fields of astronomy—it also has a great narrative. In 2011, the US National Reconnaissance Office (NRO, the agency responsible for spy satellites) revealed it had two unfinished space telescopes that it no longer wanted. Although the telescopes were designed to point down at the Earth, they would also work well pointing up at the stars. The NRO offered these obsolete telescopes to NASA, and one of them is being fitted with astronomical gear to become the Roman Space Telescope. This confirms what astronomers have long suspected—as proud as we are of our technology, the military is way ahead of us. It's a nice touch that the NRO had not just one but *two* spectacular telescopes sitting around, collecting dust, either of which would fulfill astronomers' dreams.

The Direct-Imaging Method

Eventually, we will want to build a telescope capable of *direct imaging* of Earth-like planets. If we succeed in finding and studying Earth-like planets with the Doppler, transit, astrometric, and gravitational lensing methods, it will be cause for celebration, but it will only increase our yearning to see such a planet in a direct image. In the words of Carl Sagan, we will want to see an Earth-like exoplanet not as a wiggle in a Doppler chart, or a blip in a brightness record, but as a "pale blue dot."

We are far from this goal. Direct imaging has revealed some spectacular systems, such as HR 8799, Beta Pictoris, and PDS 70, discussed in chapter 7. For each one of those systems, though, several hundred other stars were laboriously searched with the same instruments, yielding nothing. This is because our current technology, while impressive in engineering terms, is limited to detecting planets that are unusually large, hot, and far from a star. Even when a detection is made, the interpretation is often ambiguous because the planet's mass cannot be measured. Is that faint point of light really a giant planet, or is it a brown dwarf?

If we could improve on the state of the art for making high-contrast images by an order of magnitude, we could start finding giant planets analogous to Jupiter and Saturn. Another order of magnitude, and we could see exoplanetary analogs of Uranus and Neptune. Yet another order of magnitude, and we could see an Earth-like planet as a pale blue dot orbiting a nearby Sun-like star (plate 15). We could track the variation in brightness and color of the dot to learn about the planet's rotation rate and global weather patterns. We could obtain a spectrum, to learn about the contents of its atmosphere.

Achieving this goal is going to require a space telescope even mightier than the Roman Space Telescope. In 2021, a panel convened by the National Academy of Sciences recommended that building such a telescope should be the highest priority of NASA's Astrophysics Division over the next couple of decades. It's a daunting engineering challenge. It will probably involve an optical telescope with a mirror about 6 meters across, equipped with a coronagraph to suppress the starlight as much as possible while still allowing any light from planets to be focused onto the image. Astronomers and optical engineers are exploring exotic new coronagraph designs to make higher-contrast images.

Another option is a space umbrella. The idea is to build a giant opaque screen and fly it in front of the telescope, thereby preventing the star's light from entering the telescope and overwhelming the light from planets. This *starshade* would need to be in just the right location in front of the telescope to block the light from the star but not the light from the star's planets. It would function as a coronagraph located outside the telescope, instead of inside. The starshade would also need to have a precisely engineered shape. For example, a big circular disk would not work, because of the effects of diffraction. The shadow of a circular obstruction is not a crisp circle; diffraction causes the shadow to be surrounded by concentric rings of light, which would be bright enough to hide any planets.

The idea to use a starshade for detecting Earth-like exoplanets was presented in 2006 by Jonathan Arenberg, of Northrop Grumman Space Systems, and Webster Cash, then at the University of Colorado. Subsequent work has led to a conceptual design for a starshade that I find appealing not only because it's capable of achieving a contrast of ten billion to one, but also because it's beautiful. It looks like a flower with dozens of pointed petals emerging from the center (plate 16). The shapes of the petals were chosen to create the darkest possible shadow at the location of the telescope. Engineers at NASA's Jet Propulsion Laboratory have figured out how they might fold and package this flower-shaped starshade into a compact container that would fit within the nose cone of a rocket and survive the rigors of a launch. Once perched in space, the starshade would unfurl like a morning glory blooming in the sun.

As always, there's a catch. To produce a shadow with the right properties, the starshade needs to be about 25 meters across, and must be located about 10,000 kilometers in front of the telescope, quite unlike the artist's conception shown in plate 16. Two different spacecraft would need to be launched, one with the telescope and the other with the starshade. The starshade and the telescope would need to maintain their alignment to within a few meters, despite being separated by a distance comparable to the Earth's diameter. Each time we want to observe a new star, the starshade would have to propel itself to a new location thousands of kilometers away. So, the starshade would need a rocket of its own, and plenty of fuel. Nobody said finding pale blue dots would be easy.

Interstellar Objects

In addition to following the progress of the transit, Doppler, astrometric, gravitational lensing, and direct imaging methods, there are two other stories worth keeping an eye on. As of now, these are sideshows, pursued by only a small number of astronomers, but each of them has the potential to take the main stage in the coming decade.

First, there is what we might call the lazy approach to exoplanetary science. Instead of working so hard to peer at distant stars, why not let exoplanetary material come to us? I do not refer to Unidentified Flying Objects, Unidentified Aerial Phenomena, or alien visitations. I am referring to large chunks of material that are ejected from other planetary systems and randomly drift into the Solar System. Astronomers have already detected two objects that fit this description. They were recognized by virtue of their extremely high speeds. In each case, the object zoomed into the Solar System on a nearly straight line, made a sharp turn because of the gravitational attraction to the Sun, and flew off in a different direction, never to be seen again (figure 8.3). These interstellar interlopers left behind several mysteries.

The first object, discovered in 2017, was named 'Oumuamua, after a Hawaiian word meaning scout or messenger—because the object has traveled so far from home, and because the observatory that spotted it is in Hawaii. In 2019, the second object was discovered by and named after Gennadiy Borisov, an engineer and amateur astronomer. Interstellar in-

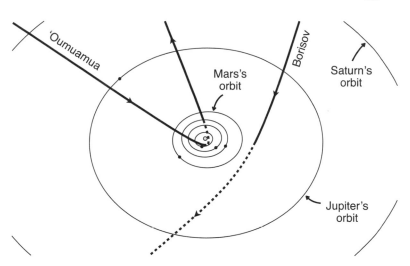

FIGURE 8.3. Trajectories of the first two interstellar objects to be discovered. Borisov was probably a comet ejected from another planetary system. The origin of 'Oumuamua is more mysterious.

terlopers such as 'Oumuamua and Borisov are probably common, given that two were found within only a few years of somewhat haphazard searching. Published statistical analyses suggest that—very roughly—there is one such object per every 10 cubic AU of space. The startling implication is that at any moment, a sphere centered on the Sun that extends to the distance of Neptune contains about 10,000 interstellar objects whizzing through the Solar System in random directions.

Borisov was discovered on its way into the Solar System, a lucky break that gave astronomers nearly a year to study it before it departed and became undetectably faint. By all accounts, Borisov resembled a comet. Comets are big dirty snowballs—kilometer-sized chunks of rocks and dust embedded in ice. The majority of the comets that appear in the night sky follow trajectories that, when traced backward, come from random directions and from distances at least 10,000 AU away from the Sun. In 1950, these facts inspired astronomer Jan Oort to hypothesize that most comets come from an enormous spherical expanse of debris surrounding the Sun, extending about a third of the way to the nearest neighboring stars. This theory is widely accepted even though kilometer-sized objects in the "Oort Cloud" would be much too small and far away to be seen through any telescope.

The debris in the Oort Cloud is probably solid material that grew from grains of dust and flecks of ice within the Sun's protoplanetary disk during the epoch of planet formation (chapter 2), before being flicked away by gravitational close encounters with Jupiter, Saturn, Uranus, and Neptune. Often, the debris would have been ejected with a high enough speed to escape the Solar System altogether, but in some cases, the debris would have remained weakly bound to the Solar System, forming the Oort Cloud. Borisov was flying way too fast to have originated in the Oort Cloud. It appears to have been a comet that was ejected from a *different* planetary system and stumbled into the Solar System after eons of drifting in darkness through the galaxy.

Comets are precious to astronomers because they are believed to be intact samples of the primitive material from which the planets were fashioned. By studying exoplanetary comets like Borisov in detail, we could see whether other stars offer a different menu of raw materials for planet-making, which might shed light on the planet formation process and help to explain the observed diversity of exoplanetary systems. Based on its brightness, Borisov was probably about half a kilometer across. As it approached the Sun, it grew a tail of water vapor, just like a comet. In fact, Borisov would easily have been mistaken for a humdrum Solar System comet, were it not for its high speed. The main puzzle posed by Borisov is that prior to its discovery, exoplanetary comets drifting through the Solar System were predicted to be so rare that detecting even one with currently available telescopes was deemed implausible.

The interpretation of the other interstellar visitor, 'Oumuamua, is murkier. It was discovered only after it passed the Sun and was heading out of the Solar System. Astronomers had only a few months to study it. Even though 'Oumuamua came closer to the Earth than Borisov, it was significantly fainter than Borisov, indicating that 'Oumuamua is smaller—maybe a tenth of a kilometer across. It appeared only as a point of light and never developed a cometary tail. Sensitive spectroscopic searches were undertaken for carbon-bearing gases such as carbon dioxide, carbon monoxide, and cyanide, which are usually seen venting from comets but were not detected coming from 'Oumuamua.

All these observations would have made sense if 'Oumuamua were more like an *asteroid* than a comet—if it were composed entirely of rock and metal rather than being a dirty snowball. There's a problem, though. An asteroid-like object would have followed the unique path dictated by the Sun's gravity and Newton's laws of motion. But when 'Oumuamua was near the Sun, its trajectory deviated slightly from the expected path as though it were being nudged outward. That's normal behavior for a comet, but not an asteroid. When a comet is warmed by the Sun, some of its icy material vaporizes and bursts from the surface to form jets of escaping gas, which act as natural rocket thrusters. Asteroids don't have thrusters because they lack large quantities of icy material. The mystery of 'Oumuamua is its dual asteroid/comet nature. How can we explain the comet-like deviations in its trajectory, while

240

also explaining why it didn't grow a tail and nobody detected any escaping gas?

Abraham (Avi) Loeb, a theoretical astrophysicist at Harvard University, promulgated the hypothesis that 'Oumuamua is an artificial object created by an advanced extraterrestrial civilization. Perhaps it was a "light sail" that uses starlight for propulsion, just as a sailboat uses the wind. It's a little-known fact, outside of physics departments, that sunlight exerts pressure on any object it strikes. The pressure is minuscule in everyday life, which is why you are not thrown to the ground when the Sun comes out from behind a cloud. But if you were floating in space while tethered to a thin, lightweight, highly reflective sheet a hundred meters on a side, the Sun's radiation pressure would exert a powerful force. Loeb and his colleague Shmuel Bialy found that 'Oumuamua's trajectory could be explained if it were a millimeter thick and made of a shiny metal, like heavy-duty aluminum foil.

Although I recommend reading Loeb's book, *Extraterrestrial*, which presents his hypothesis in detail, I'm not persuaded that 'Oumuamua must be artificial. Not because I think Loeb's calculations are incorrect; they're perfectly sound. And not because I think the possibility of alien life is ridiculous. Rather, I am unpersuaded because the alien hypothesis is too flexible. For example, 'Oumuamua's brightness rose and fell every eight hours, probably due to rotation. Asteroids are known to rotate, but why would aliens set their light sail spinning instead of keeping it facing the Sun? Loeb has suggested that it's because the sail was damaged; it's a piece of space junk.

Any mysterious finding in astrophysics—and there are many—can be "explained" by invoking suitable alien technologies and motivations. In the words of science fiction writer Arthur C. Clarke, "Any sufficiently advanced technology is indistinguishable from magic."[6] If we allow ourselves to entertain the possibility of alien technology, why not also consider the possibility that 'Oumuamua was a comet-like object that vented nitrogen or other gases that nobody looked for, instead of the carbon-bearing gases that were ruled out by the observations? Or that the venting occurred sporadically, and the gas was not present when the spectroscopists went looking for it? Or that it was a cosmic "dust bunny"—an extremely porous aggregate of microscopic dust that would have been easily pushed around by radiation pressure? Such explanations are ad hoc and unconvincing, but are they much worse than the alien hypothesis? It's only been a few years since 'Oumuamua passed through. Maybe we need to allow more time for a clever astronomer to think of a less contrived theory.

There's something else in the back of my mind. Scientists are often led astray by an observational claim that turns out to be mistaken. Whenever we have a collection of observations that together seem to rule out all reasonable hypotheses and drive us to a highly unexpected conclusion, experience shows that one of the data points often turns out be erroneous. The error only comes to light after closer examination or collection of more data. There is an apt quotation attributed to Francis Crick: "No theory should fit all the facts, because not all of the facts are true." With 'Oumuamua, it's frustrating that closer examination and collection of more data are impossible. It's too far away. To make progress, we need to find more

6. Speaking of Clarke, his 1973 novel *Rendezvous with Rama* is about a cylindrical alien spaceship coasting through the Solar System. Initial reports of 'Oumuamua suggested it was cylindrical, too—an entertaining connection between science and science fiction—but subsequent analysis showed that 'Oumuamua was probably more disk-shaped than cylindrical.

THE WORLDS TO COME

interstellar objects. On this, Loeb and I, and probably every other astronomer, agree.

We will probably not need to wait long, thanks to a new observatory scheduled to begin operations in 2024. Located in northern Chile, the Vera Rubin Observatory will use a visible-light telescope and the world's largest-ever astronomical camera (weighing three tons) to scan the sky for anything that moves or changes brightness. The telescope's main mirror is 8.4 meters in diameter, plenty large enough to see faint objects like 'Oumuamua. The observatory is being built for many reasons, including the study of distant galaxies and supernovae, but it will also be an excellent tool for finding interstellar asteroids and comets, at an expected rate of a few per year.

Once we've observed many interstellar objects, we'll be able to compare the range of their colors, compositions, and sizes with the ranges seen in the Solar System. We might gain more clues to the mystery of 'Oumuamua by finding similar objects and performing more intensive observations. Several teams of astronomers and engineers are even making plans for rockets that would be able to catch up with an interstellar object and take close-up pictures before it departs the Solar System. Close-up pictures would tell us whether the object is an exo-asteroid, an exo-comet, or an exo-sailboat.

Planet 9

Another story worth following, which could fizzle out or become front-page news, pertains to super-Earths and mini-Neptunes. As discussed in chapter 5, these medium-sized

planets are commonly found orbiting Sun-like stars, although
the Solar System itself does not have one.

Or does it?

In 2016, Caltech astronomers Michael Brown and Konstantin Batygin, building on earlier work by Chadwick Trujillo
and Scott Sheppard, claimed to have found evidence for a
super-Earth inside the Solar System. The hypothetical planet
is too far and faint to have been detected directly but, they
argued, the planet's fingerprints can be found on the orbits of
known objects. Specifically, Batygin and Brown blamed the
unseen planet for some seemingly strange coincidences in the
orbital parameters of some of the known objects in the Solar
System's Kuiper Belt.

The Kuiper Belt is a collection of small icy-and-rocky
bodies found beyond the orbit of Neptune, analogous to the
asteroid belt that exists between the orbits of Mars and Jupiter. Pluto is the Kuiper Belt's most famous inhabitant, but
thousands of others are known, and a minority of the known
comets come from there, too. Almost all the known Kuiper
Belt Objects, or KBOs, are currently within about 50 AU of
the Sun, making them far away but not nearly so far as the
Oort Cloud. A few dozen of the known KBOs have extremely eccentric orbits that extend to hundreds or thousands
of AU from the Sun. The origin of these relatively far-ranging
KBOs is unknown. Some of them probably started out on
orbits that stayed closer to the Sun, like the rest of the KBOs,
but were scattered outward by the gravitational effects of encounters with Neptune. A few of them, though, are on orbits
that *always* stay far from Neptune. Their orbits appear to be
pristine, never having been affected by Neptune or the other
planets.

Brown and Batygin focused their attention on six KBOs with the most pristine orbits, seemingly detached from the rest of the Solar System. After poring over the data, they became convinced that *something* massive must have altered their orbits, after all, because all six orbits are geometrically similar. The orbits are roughly aligned with each other, and their average orbital plane is tilted away from the plane defined by the Solar System's eight planets. Furthermore, the long axes of all six ellipses are all pointing in roughly the same quadrant of the Solar System, even though we would expect them to be pointing in random directions.

A strategically placed distant planet can account for these coincidences. The planet's gravity would destabilize many of the possible orbits for a small body in the vicinity. Besides the obvious danger of crashing into the planet, a KBO runs the risk of being ejected from the Solar System after a close encounter with the planet. The planet's gravity would also cause more gradual changes in the KBO's orbit—and sometimes, those gradual changes would send the KBO toward Neptune to have a fatal close encounter. According to Batygin's calculations, the only KBOs that would stay safe are those on elliptical orbits oriented with their long axes pointing in the opposite direction as the long axis of the planet's orbit (figure 8.4).

In short, Batygin and Brown think that the six special KBOs are the survivors from a larger collection of objects, most of which were destroyed or ejected due to the gravitational effects of an unseen massive planet. Furthermore, Batygin and Brown were able to explain the overall tilt of the six KBO orbits by supposing that the unseen planet's orbit is inclined with respect to the rest of the Solar System. They called their idea the *Planet 9 hypothesis*, with a nod to the campy 1957

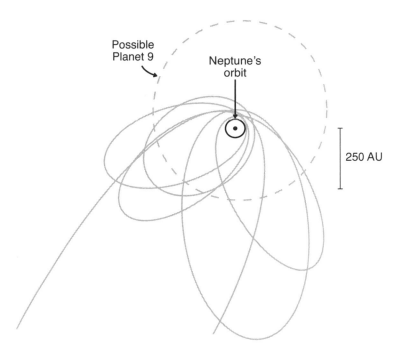

FIGURE 8.4. Orbits of some of the most distant known Kuiper Belt Objects. The unexpected similarities in the orientations of the major axes and orbital planes could be due to the gravitational effects of an undiscovered outer planet (dashed ellipse).

film *Plan 9 from Outer Space*. For it to have committed the orbital crimes of which it is accused, Planet 9 must have a mass of about 6 $M_\oplus$, an elliptical orbit that ranges between about 300 AU and 450 AU, and an orbital tilt of about 15° relative to the Earth's orbital plane.

This is a textbook example of the scientific method: notice an anomaly, construct a hypothesis, and make predictions. It harkens back to Le Verrier's successful prediction of the existence of Neptune, based on anomalies in the motion of Uranus. An important factor in the credibility of the Planet 9 hypothesis is the renown in which Batygin and Brown are held by the astronomical community. Brown is the astronomer who started the Pluto wars in 2005 by discovering a KBO rivaling Pluto in size, thereby forcing the question of whether Pluto should be considered a planet. Since their original proposal in 2016, Batygin and Brown have accumulated more evidence for their hypothesis. They identified five more KBOs that seem to have been affected by Planet Nine, for a total of 11, while also arguing that Planet 9 could be responsible for a separately puzzling category of Kuiper Belt Objects called the high-inclination Centaurs. Their hypothesis ties together several loose ends into a satisfying knot.

Now, for a dash of skepticism. The statistical significance of the coincidences in the orbital data is not overwhelming. The orbital planes of the KBOs in question are not perfectly aligned; they are aligned to within 20°. The orientations of the ellipses span a range of 110° out of the maximum possible range of 360°, and that's after removing one outlying data point. So, while the patterns appear to be present, they don't leap off the page. In 2016, Brown and Batygin estimated the probability that they were being fooled by a statistical fluke to be only 0.007%, but in 2021, using more sophisticated calculation, they revised this false-positive probability upward to 0.4%. It's a hairy calculation because the KBOs were not all discovered in the same way. They were found by different teams who were looking in different parts of the sky, raising the question of whether the similarities in the orbital parameters of the KBOs are a consequence of the uneven coverage of the sky.

I also need to fess up about the historical precedent. We regale our astronomy students and astronomy book-readers with the story of Le Verrier and Neptune because it was a triumph of the human mind, but in fact there have been many more historical episodes in which astronomers made false predictions of the existence of a distant planet. In 1880, based on the similarity of the orbital parameters of some comets, astronomer George Forbes became convinced that there were two massive planets beyond Neptune. The patterns turned out to be meaningless. In 1915, Percival Lowell saw anomalies in the motion of Neptune that he attributed to the tug of a massive planet. This turned out to be due to an error in the assumed mass of Neptune.[7] Throughout the first part of the twentieth century, William Pickering postulated planet O, then planet P, and eventually Q, R, S, T, and U. None of them was real. Even in more recent years, following the discovery of the first KBO in 1992, there have been several unsuccessful attempts to predict the existence of a new planet based on various features of the Kuiper Belt.

The Planet 9 hypothesis is frustrating for another reason. We should be able to find Planet 9 by scanning the sky with telescopes large enough to detect a super-Earth located as far away as 450 AU. But we don't know where to look. The Planet 9 hypothesis specifies the planet's mass and orbital characteristics but not where the planet is located along the orbit. It could be anywhere within a wide stripe that runs all the way around the celestial sphere. In that sense, the hypothesis is not as predictive as Le Verrier's calculations that guided astronomers to Neptune. Two places where it would be especially difficult to detect Planet 9 are the intersection points between the planet's hypothetical orbit and the luminous band of the

---

7. Lowell's work inspired Clyde Tombaugh to search for the predicted planet, resulting in the discovery of Pluto, but the discovery was a lucky coincidence. Pluto's mass is 3,000 times smaller than predicted by Lowell.

Milky Way. The throngs of distant stars in the background make it harder to spot a faint planet in the foreground.

Having aired my skepticism and frustration, I would like to return to a brighter mood. Brown, Batygin, and other astronomers have been successful in convincing Telescope Time Allocation Committees to let them keep searching for Planet 9. And remember the Vera Rubin Observatory, which will soon begin scanning the sky from its perch on a mountain in northern Chile? The data from Rubin will allow for a comprehensive search for Planet 9 in the southern sky. Given all these efforts, if Planet 9 exists and is not diabolically using the densest part of the Milky Way to hide itself, there's a good chance it will be found by 2030.

Finding a super-Earth in the Solar System would be a spectacular once-per-century type of astronomical discovery. Imagine having another entire world to explore, different from all the other planets. With targeted telescopic observations, we could learn the details of its orbit and atmosphere. We might even be able to send a spacecraft to take close-up pictures, although that would require patience, given the expected distance to Planet 9. Even at the speed of NASA's *New Horizons* spacecraft, which made a trip to Pluto in a decade, it would take a spacecraft a century to reach Planet 9.

The mere existence of Planet 9 would call into question our understanding of the history of the Solar System. How could such a super-Earth have formed way out there? At such large distances, by our current understanding, there should not have been enough material in the protoplanetary disk to form a planet more massive than Earth. Nor would there have been enough time for dust and rocks to congeal into a planet, given the cold conditions and slow orbital speeds that prevail so far from the Sun. Maybe a planet that formed closer to the Sun was flicked outward by Jupiter or one of the other giant planets, and then its orbit was permanently widened due to a close encounter with another star. Improbable though that might seem, exoplanetary science has given us a new appreciation of all the possibilities for tangled gravitational interactions and drastic planetary rearrangements.

A more adventurous theory for the formation of Planet 9 (should it exist) is inspired by Triton, the largest moon of Neptune. Triton's orbit is nothing like the orderly, well-aligned, circular orbits of Jupiter's large moons. Triton is in an unusually wide orbit that revolves backwards and is tilted by 23 degrees with respect to Neptune's equator. The explanation is that Triton did not form along with Neptune. Instead, Triton was an errant KBO that was accidentally captured by Neptune in a gravitational encounter. All the giant planets have irregular satellites like Triton.

With that in mind, could the Sun have captured an errant planet from another star? When the Sun was born, it was probably part of a cluster of hundreds or thousands of stars, each descending from a different fragment of a massive cloud of interstellar gas. The Sun and its brethren remained close together during their youth, before the stars' random motions caused them to drift apart over hundreds of millions of years. In those early crowded conditions, a close encounter between two stars was more likely than it is today. While it is still unlikely, in principle those encounters could have plucked a planet from one star and attached it to another, leaving the planet in a weakly bound orbit with no geometrical kinship with the star's preexisting planets.

How wonderful it would be if Planet 9 were an exoplanet, right in our backyard!

AFTERWORD

Imagine waking up one morning and reading the headline: "Astronomers discover life on an exoplanet."You'd be stunned. It would take a long time to absorb the knowledge that there are other creatures in the galaxy, living on their own planet, with their own biology—maybe even their own civilizations, their own purposes, accomplishments, failures, ambitions, and fears. All the other news articles about the shenanigans and skirmishes on Earth would seem trivial by comparison. The discovery of life elsewhere in the universe would be transformative, dividing history into Before and After the realization that we are not alone.

With a second example of life in the universe, we could finally make decisive progress on the question of how life begins. What are the necessary and sufficient conditions for chemistry to become biology? The transformation through natural spontaneous processes from simple molecules and chemical reactions into complex and self-reproducing protein-manufacturing equipment, and then into goal-seeking creatures with emotions, intelligence, problem-solving ability, and the capacity to discover all of this . . . is so mind-blowing and awe-inducing that it gives me chills every time I sit down and really think about it.

If we find intelligent life-forms, we will have an intense desire to communicate. What might their civilization teach ours about the workings of the universe, about how to overcome the challenges of long-term existence, about the meaning of it all? There would be implications for science, technology, politics, religion, and, well, everything. So, it's no wonder that most accounts of exoplanetary science for the general public focus on the search for life on other planets.

In this book, I wanted to stay close to the data. The search for life has made only a few cameo appearances, interspersed in discussions of gravity, radiation, the physics of planet formation, and astronomical instrumentation. We know much more about planets today than we did 30 years ago. We know planets are commonplace; we know about varieties of planets that are not represented in the Solar System; we know the universe has plenty of potential habitats for life. Apart from these facts, though, exoplanet astronomers can make few, if any, definitive statements relevant to extraterrestrial life.

Searching for life is a quest of unknown difficulty and duration. We have no good evidence for life anywhere else in the universe, besides Earth. As recently as the 1950s, astronomers could realistically hope that spacecraft sent to Mars would reveal creatures crawling or hopping around the surface. Today, our reconnaissance of the Solar System has ruled out widespread macroscopic life-forms on the surfaces of the other terrestrial planets. The search for alien life inside the Solar System will continue, but in remote places, by digging into the Martian soil and exploring the underground oceans of Jupiter's moon Europa.

To search *exoplanets* for signs of life might sound far more difficult than searching the planets of the Solar System—or even impossible, given the vast distances. If searching the Solar

System is like searching your bedroom to find your phone, then searching the space within 20 light-years is like trying to find a phone that could be anywhere in North America. Paradoxically, though, detecting life on exoplanets could turn out to be *easier* than detecting it in the crevices of the Solar System. There are only seven other planets in the Solar System, but there are countless exoplanets. Even if the other planets in the Solar System lack obvious planet-wide signs of life, some exoplanets could be teeming with life in their atmospheres and on their surfaces. Intelligent life-forms might even be broadcasting their existence with messages beamed into space. Returning to the phone analogy, it's now a simple matter to find a misplaced phone anywhere in North America, because phones broadcast their own locations.

In chapter 7, I described the "biosignature gas" approach to searching for signs of life on an exoplanet. Obtain the exoplanet's spectrum using either the transit method or the direct-imaging method. Search the spectrum for absorption by a gas, or combination of gases, that is difficult to explain by natural processes except for the activity of creatures trying to make a living. For example, we can look for absorption by oxygen gas, which exists in our atmosphere only because of photosynthetic organisms.

This idea took flight in 1993 when Carl Sagan and a few of his colleagues published a charming paper in which they pretended to be alien astronomers trying to detect life on Earth. They analyzed data from the Galileo spacecraft, an unmanned probe that had been sent by NASA to explore Jupiter. On its way there, Galileo observed the Earth with a few of its astronomical instruments. Sagan's team looked at Earth's spectrum and found "evidence of abundant gaseous oxygen, a widely distributed surface pigment with a sharp absorption edge in the red part of the visible spectrum, and atmospheric methane in extreme thermodynamic disequilibrium; together, these are strongly suggestive of life on Earth." Methane is emitted when organic matter decays, or, famously, when organic beings flatulate. If all the methane-producing creatures suddenly vanished, within a few decades most of the methane in the atmosphere would be gone, having reacted with oxygen to form carbon dioxide and water. That's why Sagan and his colleagues, posing as aliens, argued that the simultaneous presence of oxygen and methane was indicative of life.

As we've seen, obtaining the spectrum of an Earth-like exoplanet will probably require a new generation of space telescopes. Even after achieving this feat, how long will it take to interpret the spectrum? Suppose we find oxygen and methane—how certain will we be that life is the only possible explanation? Could an unrecognized type of vulcanism or atmospheric chemistry be fooling us? In 2004, astronomers detected traces of methane gas in Mars's atmosphere, fueling speculation about methane-exhaling bacteria buried in the Martian soil. After this finding caught the attention of the broader community of planetary scientists, some of them provided more mundane explanations for Martian methane. For example, even though there are no active Martian volcanoes today, the methane could have erupted from ancient volcanoes and gotten trapped within the ice on Mars's surface—a type of ice called clathrates, in which water molecules form a cage around other molecules like methane. In this hypothesis, whenever clathrates melt, the atmosphere gets a dose of methane.

Another interesting development took place in 2020, when a group of astronomers led by Jane Greaves, of Cardiff University, announced the detection of phosphine (PH_3), a possible biosignature gas, in Venus's atmosphere. There's still controversy over whether the detection is genuine. Some experts have called into question the team's techniques for analyzing the data, and others have pointed out that the spectral feature that appears to be PH_3 might be from the nonbiological gas SO_2. If the detection withstands further scrutiny, it will be interesting to see how long it takes before planetary science journals publish articles about nonbiological processes that produce phosphine.

Given the uncertainties so far in the data from Venus and Mars, our nearest planetary neighbors, I worry that the interpretation of analogous data from distant exoplanets will *always* be ambiguous. A recent review article about exoplanet atmospheres called the discovery of extraterrestrial life using biosignature gases the "holy grail of exoplanetary science." Will this goal be just as elusive for astronomers as it was for King Arthur?

I think the only way we will ever be certain of life on exoplanets is if we detect signs of *intelligence*, such as patterns in the radiation from a planet that bear information, or perhaps a deliberate message sent across the void, like the radio message received by Jodie Foster's character in the 1997 film *Contact*.

In 1974, a team of astronomers led by Frank Drake broadcasted a radio message to the hundred thousand stars in the Hercules Globular Cluster, located 25,000 light-years away. The 210-byte message, in a binary code, started by counting the numbers from 1 to 10 before proceeding to more complex numerical constructions conveying the chemical formula of DNA and a chart of the Solar System. If instead of *transmitting* we had *detected* such a message from the Hercules Cluster, we would know that we are not alone—or at least that we were not alone 25,000 years ago when the message was sent. Even if we don't receive a deliberate message, we might be able to detect inadvertent leakage of radio signals produced for other purposes, such as the aliens' communication network or radar defense system. If today's most powerful radar transmitters were sending signals at maximum intensity to the Earth from a planet orbiting Proxima Centauri, we would be able to detect the signals with existing radio telescopes.

The most important role of biosignature-gas investigations might be to produce a short list of planets that are worth intensively searching for intelligent broadcasts. We can lavish much more observing time and other resources on the planets where something funny seems to be going on in the atmosphere. By focusing our attention on promising systems, maybe we will accelerate the search.

Sometimes, the search for life on exoplanets feels like wishful thinking. It seems improbable that we would be able to deduce the existence of life based on hints in a planet's atmosphere, or anticipate the characteristics of an extraterrestrial message well enough to design an effective search program. Even the concept of the "habitable zone" might be a misleading oversimplification. Is liquid water truly essential for life? Does it guarantee the eventual emergence of life? We don't know.

Given the lack of hard evidence, there's room for both optimism and pessimism about the prospects for a universe teeming with life. Pessimists can point to the evidence that all known life-forms share a common ancestry, based on the universality of key biochemical cycles and the genetic code. This is despite the billions of years that were available for life to arise again and again—suggesting genesis is a very rare event. Optimists can reply that independent strains of life might have arisen, before eventually going extinct and leaving no trace in the fossil record. Optimists can also point to geological evidence for life that existed 3.8 billion years ago, only a few hundred million years after the Earth had cooled down enough for ocean-based life to be physically plausible. If true, this suggests genesis is rapid once the right conditions are met.

Astrobiologists write papers weighing these shreds of evidence and struggling with the "Fermi paradox," which goes as follows. If there are widespread and long-lasting alien civilizations in the galaxy, then where are they? Shouldn't they have the technological capability to travel between stars, or at least make their existence obvious? Does the fact that we have not been visited by aliens tell us that civilizations tend to die out before they develop interstellar spacefaring technology? These questions are fascinating to ponder, but to me, this genre of papers carries a whiff of scholasticism. How many aliens can dance on the head of a pin?

Sometimes, though, when I'm in a different mood—when I'm in the same frame of mind as I was when I was a little boy on vacation in the Grand Canyon, staring up at the stars—the search for extraterrestrial intelligence seems like one of our civilization's most forward-looking activities. We should be thankful there are at least a few special people on Earth who believe so strongly in the transcendent importance of finding life elsewhere in the universe that they are willing to devote their careers to such a speculative endeavor.

I'm also impressed and encouraged by the sheer numbers of planets and the vast spatial scales that modern astronomy has uncovered. The starry night sky is inspiring, but it gives us only the slightest inkling of the true scale of the universe. Imagine viewing the Solar System from above, and allow your mind to zoom out. Watch the Sun and planets recede until the

Sun is one of a thousand points of light. These are the stars in our neighborhood, close enough for us to have a conversation with any other residents. We could send a message and receive a reply within a single human lifetime.

Zoom further until your view encompasses hundreds of millions of stars. Almost all the known exoplanetary systems are within this region of our galaxy. It's getting hard to distinguish them individually; the stars have seemingly merged to form a luminous fluid, the Milky Way. With a little more zooming out, you can admire the vast circular disk and spiral patterns of the Milky Way, which stretch out from a dense concentration of stars at the center. A few smaller galaxies and clusters of stars come into the picture, and over there in the distance is Andromeda, the nearest spiral galaxy, the home of another hundred billion stars.

Expand your mental scene another order of magnitude. The Milky Way and Andromeda are now surrounded by other galaxies with a variety of shapes and sizes. At first, they appear clustered together, with groups ranging from a dozen to thousands, separated by voids of empty space. But once we expand our horizon another few more orders of magnitude, we see that the universe is a formless sea of galaxies. As far as we can tell, the sea has no shores.

From this perspective, planets seem utterly insignificant. They are specks of dirt that collect around stars, like lint in a navel. Even the stars are infinitesimal. A star is smaller than a galaxy by the same factor that an atom is smaller than a human being. Once we consider scales of billions of light-years, entire galaxies play the role of atoms.

But there's another way to think about the situation. Think about the process of planet formation: those grains of dust that stick together to form pebbles, which conglomerate into rocks and boulders, which attract each other to form planets, which accrete gas, and undergo collisions and rearrangements. Isn't it baroque and magnificent? There are so many seemingly unlikely and poorly understood steps. It feels like a lucky break that planets exist at all. Nowhere else in the known universe, apart from planets, do we find the conditions suitable for liquids to accumulate and for complex chemical reactions to take place over billions of years. Surely, given the unfathomable number of galaxies, stars, and planets in the universe, and the billions of years that have elapsed, every possible chemical reaction has been repeated many times in many places, including those that spawn life. The only question is how far we need to look.

AFTERWORD

ACKNOWLEDGMENTS

I'm very grateful to John Richards, Harris Richter, Adam Lewis, Irwin Shapiro, Dalton Conley, Sophie Winn, Kate Ivshina, Konstantin Batygin, Joel Shapiro, my editor Abigail Johnson, and two anonymous reviewers, for constructive criticism of the manuscript. Some of the material in the book is related to lectures I recorded for the Teaching Company (Wondrium), where Susan Dyer provided expert assistance. Michael Lemonick helped with an early version of chapter 3 that was published in 2019 on the *Scientific American* website. Sarah Millholland, Morgan MacLeod, Sean Andrews, Konstantin Batygin, and Gáspár Bakos assisted with some of the illustrations. The Institute for Advanced Study in Princeton provided ideal working conditions during the year this book was written.

I cannot adequately convey my everlasting gratitude to my parents, Martin and Barbara Winn, my companion star, Lara, and our circumbinary children, Sam and Sophie.

FURTHER READING

Below are suggestions for further reading associated with each chapter. In general, I selected books and articles intended for a broad audience. Those with training in physics and astronomy who want to delve more deeply may be interested in these resources:

Seager, S., editor, *Exoplanets*, University of Arizona Press (2010).

Perryman, M., *The Exoplanet Handbook*, 2nd ed., Cambridge University Press (2018).

Winn, J. and Fabrycky, D., "The Occurrence and Architecture of Exoplanetary Systems," *Annual Reviews in Astronomy and Astrophysics*, Vol. 53, no. 49 (2015). This technical article covers some of the same ground as this book and includes citations to the original literature.

The most comprehensive exoplanet catalog is the NASA Exoplanet Archive (https://exoplanetarchive.ipac.caltech.edu/), where you can search for all the planets described in this book by name (e.g., WASP-12b, HD 80606b, Beta Pictoris).

You can search the professional astronomy literature for your favorite exoplanet or for topics discussed in this book by using the NASA Astrophysics Data System (https://ui.adsabs.harvard.edu/). For example, try the following searches:
author:"Winn" and object:"WASP-12"
abstract:("Beta Pictoris", "comets")

Chapter 1

For learning about the night sky, the best book is H. A. Rey, *The Stars: A New Way to See Them*, Clarion Books (2016); and the best magazine is *Sky & Telescope*, published by the American Astronomical Society.

For more on the quest to measure the distances to the stars, see A.W. Hirshfeld, *Parallax: The Race to Measure the Cosmos*, Henry Holt and Company (2002).

For more on Yuri Milner and his motivation for funding Breakthrough Starshot and other space-related initiatives, see his online manifesto (https://yurimilner manifesto.org/).

Chapter 2

For more on the development of calculus and its application to planetary motion, see S. Strogatz, *Infinite Powers: How Calculus Reveals the Secrets of the Universe*, Houghton Mifflin Harcourt (2019).

Proofs of Kepler's laws using calculus are provided in many advanced undergraduate astrophysics textbooks, such as B. W. Carroll and D. A. Ostlie, *An Introduction to Modern Astrophysics*, 2nd ed., Cambridge University Press (2017).

For a proof of Kepler's laws without calculus, see David Goodstein and Judith R. Goodstein, *Feynman's Lost Lecture*, W. W. Norton (2009); or Grant Sanderson's video version at https://www.youtube.com/watch?v=xdIjYBtnvZU.

For a history of planet formation theory up until the Apollo age, see I. Williams and A. Cremin, "A Survey of Theories Relating to the Origin of the Solar System," *Quarterly Journal of the Royal Astronomical Society*, Vol. 9 (1968): 40, available at https://ui.adsabs.harvard.edu/abs/1968QJRAS . . . 9 . . . 40W /abstract.

Chapter 3

Reminiscences written by many exoplanet pioneers can be found in a special edition of *New Astronomy Reviews*, Vol. 56, 1(2012); as well as R. P. Butler, "A Brief Personal History of Exoplanets," at https://palereddot.org/a-brief -personal-history-of-exoplanets-by-paul-butler.

Chapter 4

For more technical information about hot Jupiters, see R. Dawson and J. Johnson, "The Origin of Hot Jupiters," *Annual Reviews of Astronomy & Astrophysics*, Vol. 56 (2018): 175; and J. Fortney, R. Dawson, and T. D. Komacek, "Hot Jupiters: Origins, Structure, Atmospheres," *Journal of Geophysical Research: Planets*, Vol. 126, 3, article i.d. e06629 (2021).

Chapter 5

For more on the Kepler mission, see W. Borucki, "Kepler Mission: Development and Overview," *Reports on Progress in Physics*, Vol. 79, article i.d. 036901 (2016).

For more on the habitable zone, see J. Kasting, *How to Find a Habitable Planet*, Princeton University Press (2012).

For more on TESS, see G. Ricker, "Transiting Exoplanet Survey Satellite," *Journal of Astronomical Telescopes, Instruments, and Systems*, Vol. 1, i.d. 014003 (2015).

Chapter 6

The quote about Newton's thoughts regarding gravitational dynamics is from M. Hoskin, "Gravity and Light in the Newtonian Universe of Stars," *Journal for the History of Astronomy*, Vol. 39 (2008): 251.

To play around with planetary dynamics yourself, enjoy Stefano Meschiari's online game SuperPlanetCrash (www.stefanom.org/spc).

For more on the discovery of Neptune, see W. Sheehan, N. Kollerstrom, and C. Waff, "The Case of the Pilfered Planet," *Scientific American*, Vol. 291, no. 6 (2004): 92; and T. Standage, *The Neptune File: A Story of Astronomical Rivalry and the Pioneers of Planet Hunting*, Walker Books (2020).

Chapter 7

For more on stellar evolution, white dwarfs, neutron stars, gravitational lensing, and other astrophysical wonders, see N. deGrasse Tyson, M. Strauss, and R. Gott, *Welcome to the Universe*, Princeton University Press (2016).

Chapter 8

For more on 'Oumuamua, see A. Loeb, *Extraterrestrial: The First Sign of Intelligent Life beyond Earth*, Mariner Books (2021); and a more technical article by the 'Oumuamua ISIS team, "The Natural History of 'Oumuamua," *Nature Astronomy*, Vol. 3 (2019): 594.

For more on Planet 9, see K. Batygin, F. Adams, M. Brown, and J. Becker, "The Planet 9 Hypothesis," *Physics Reports*, Vol. 805 (2019): 1. Although this is a technical article, the first two sections about the motivation and the general principles are more accessible.

FURTHER READING

For more on Pluto and the Kuiper Belt, see M. Brown, *How I Killed Pluto and Why It Had It Coming*, Random House (2012).

For more on failed attempts to predict planets, see M. Grosser, "The Search for a Planet beyond Neptune," *Isis*, Vol. 55 (1964): 163.

Afterword

For more on the search for extraterrestrial life, see J. Al-Khalili, *Aliens: The World's Leading Scientists on the Search for Extraterrestrial Life*, Picador (2017); S. Shostak, *Confessions of an Alien Hunter: A Scientist's Search for Extraterrestrial Intelligence*, National Geographic (2009); and W. Roush, *Extraterrestrials*, MIT Press (2020).

For Carl Sagan's paper about trying to detect life on Earth, see Sagan et al., "A Search for Life on Earth from the *Galileo* Spacecraft," *Nature*, Vol. 365 (1993): 715.

FURTHER READING

GENERAL INDEX

GENERAL INDEX

INDEX OF STARS AND EXOPLANETS

ABOUT THE AUTHOR

Joshua Winn is composed of carbon, nitrogen, oxygen, sulfur, and iron atoms forged inside massive stars, as well as hydrogen from the early universe and trace elements from colliding neutron stars. He was on the physics faculty of the Massachusetts Institute of Technology for a decade before moving to Princeton University in 2016, where he is a professor of astrophysics. He has written for *The Economist* and recorded two lecture series for the Teaching Company (Wondrium), *The Search for Exoplanets* and *Introduction to Astrophysics*.